Lyubov Grigorenko

# Quality of tap and pretreated drinking water in rural taxa

Lyubov Grigorenko

# Quality of tap and pretreated drinking water in rural taxa

ScienciaScripts

**Imprint**
Any brand names and product names mentioned in this book are subject to trademark, brand or patent protection and are trademarks or registered trademarks of their respective holders. The use of brand names, product names, common names, trade names, product descriptions etc. even without a particular marking in this work is in no way to be construed to mean that such names may be regarded as unrestricted in respect of trademark and brand protection legislation and could thus be used by anyone.

Cover image: www.ingimage.com

This book is a translation from the original published under ISBN 978-3-659-82326-8.

Publisher:
Sciencia Scripts
is a trademark of
Dodo Books Indian Ocean Ltd. and OmniScriptum S.R.L publishing group

120 High Road, East Finchley, London, N2 9ED, United Kingdom
Str. Armeneasca 28/1, office 1, Chisinau MD-2012, Republic of Moldova, Europe
Printed at: see last page
**ISBN: 978-620-8-20038-1**

*Reviewers:*

**Buryak L.I.** - Doctor of Medical Science, Professor of the Department of Hygiene and Ecology "Dnepropetrovsk Medical Academy of the Ministry of Health of Ukraine", Academician of the Academy of Sciences of Ukraine, scientific director of the laboratory N-VTC "Hygienist" and research laboratory N-VTC "Expertise". Author of more than 300 scientific works: including 2 monographs; 5 inventions; 35 proposals.

**Shevchenko I.N.** - Candidate of Medical Sciences, Associate Professor, First Vice-Rector of "Dnepropetrovsk Medical Institute of Traditional and Non-traditional Medicine". Author of more than 150 scientific works.

CONTENTS.

# TABLE OF CONTENTS

**Relevance.** Analysis of the current situation in Ukraine in the sphere of drinking water supply, quality of drinking water and sanitary condition of water supply sources indicates a real danger of water factor for human health [1]. Negative trends in providing the population with drinking water of guaranteed quality have been accumulating for many decades and now in some regions of Ukraine have reached a critical state [2].

Dnepropetrovsk region is one of the largest in Ukraine in terms of contamination of water supply sources. According to the results of numerous studies, it was found that in a number of rural areas of Dnepropetrovsk region the quality of drinking water from surface water bodies does not meet sanitary requirements in more than 60% of samples for physicochemical and more than 10% of samples for bacteriological indicators [3].

However, the problem of drinking water supply quality in rural areas is not given due attention by domestic scientists. Drinking water from decentralized water supply sources in most rural areas of Ukraine does not meet the requirements of hygienic standards in terms of mineral composition: total hardness, salt content, nitrogen compounds, iron, manganese, the content of which is 2-10 times higher than MPC. But scientists do not always associate this with anthropogenic pollution of water supply sources, but with regional natural peculiarities of interlayers of soil in which water is formed [3].

Since the vast majority of scientific research is focused on the study of hygienic state of water supply of urban population, especially in industrial regions of Ukraine [4, 5, 6], the need for such studies in rural areas becomes even more argued. In this regard, the issue of studying the chemical composition of drinking water in rural areas is relevant.

Since the times of the former USSR, Ukraine has retained the practice of granting temporary permits for the use of tap water of non-standard quality in terms of mineral composition. About 4.6 million people in 160 cities and 100 urban-type settlements in 25 regions of Ukraine receive drinking water from underground water supply sources with deviations from the normative requirements [7]. However, on the territory of Dnepropetrovsk oblast with a total population of 3.4 million inhabitants - the rural population is 609365 inhabitants - the study of the chemical composition of drinking water in rural areas has not been conducted during the last decade.

In the complex impact of various environmental factors on the state of public health, a significant contribution is made by drinking water, which can cause infectious and non-infectious morbidity [8]. It is well known that non-compliance of drinking water quality with regulatory requirements is one of the reasons for the spread of diseases of non-infectious etiology: dental caries or dental fluorosis (deficiency or excess of fluoride); water-nitrate methemoglobinemia (excess of nitrates in water); urolithiasis or cholelithiasis (excess of mineral salts in water); endemic goiter (deficiency of iodine in water); cardiovascular diseases (soft or hard water) [9].

Works of domestic scientists - hygienists for the last 10 years allowed to predict dangerous consequences of active migration of heavy metals (HM) into life-supporting environments and to form their negative impact on the health of the population of residential areas of industrial cities [10]. It is proved that during the last 20 years in the air of industrial cities of Ukraine there is a gradual decrease in the content of HMs in the air, but a significant increase in their content in water and food products, which correlates with the rate of internal contamination of the organism of the inhabitants of industrial cities [11]. Therefore, the problem of studying chemical pollution in the systems of centralized and decentralized water supply in rural settlements is relevant.

A multi-year study by American scientists in rural areas of selected U.S. states, conducted during 1971-2006, identified etiologic factors in 48 cases of waterborne disease outbreaks that occurred in 24 states. Of these 48 outbreaks, 36 were associated with insufficiently treated drinking water from groundwater sources, which contributed to the emergence of infectious diseases among the adult population: 4128 people became ill and 3 people died [12].

Detailed analysis of the causes of waterborne disease outbreaks showed that 21 outbreaks (58.3%) were associated with E. coli bacteria, 5 outbreaks (13.9%) were of viral origin, 3 outbreaks (8.3%) were caused by parasites, 1 outbreak (2.8%) was associated with chemical contamination of drinking water from wells, 1 outbreak (2.8%) was due to simultaneous contamination of groundwater sources with bacteria and viruses, 1 outbreak (2.8%) was due to simultaneous contamination of drinking water with bacteria and parasites, and 4 outbreaks (11.1%) were of undetermined etiology. Among the 36 aquatic outbreaks in adults in selected US states, 22 outbreaks (61.1 %) of acute gastrointestinal diseases, 12 outbreaks (33.3 %) of acute enterovirus diseases, and 1 outbreak (2.8 %) of hepatitis A were reported [13]. The main causes of these waterborne disease outbreaks within the United States are considered by epidemiologists from the Center for Disease Control to be deficiencies associated with the consumption of inadequately treated drinking water from underground water supplies. A total of 21 (59.5 %) cases of aquatic outbreaks have been reported, with the major deficiencies including: 13 (61.9 %) cases are related to untreated drinking water from groundwater supply sources, 6 (28.6 %) to the drinking water treatment system, 1 (4.8 %) to the distribution system of pretreated drinking water, and 1 (4.8 %) to the distribution network [14].

No outbreaks were detected in the treatment of surface water supplies. More than 50% of groundwater supplies in rural areas of the United States have caused waterborne disease outbreaks associated with untreated or inadequately treated groundwater supplies over a 35-year period (1997 to 2006), hence groundwater contamination remains a pressing hygiene problem [15]. Therefore, U.S. public health agencies are focusing on the identified causes of disease especially among rural populations, sanitation of wells and drinking water sources, and sanitation of rural wells to protect the population from bacterial and viral pathogens [16].

According to the literature [ 17] it is established that the main role of influence on the health of the population is played by such risk factors as "lifestyle", unfavorable demographic situation, irrational nutrition, harmful working conditions and the like. The share of influence of these factors on health is 49-53%, the share of influence of genetic factors is 18-22%, medical factors is 8-10%, and the influence of environmental factors on health is 17-20% [18]. Consequently, when addressing the issue of environmental pollution danger for the health of rural population, it should be taken into account that harmful factors can affect not only by inhalation, but also orally - through drinking water and food [19, 20, 21]. This is especially important for substances that are widespread and easily included in biological chains: "soil - ground and surface water - plants - animals - humans". These include mainly heavy metals, persistent organic nitrogen-containing compounds and other xenobiotics [22, 23, 24].

According to the UN, currently 1.1 billion of the world's population do not have access to quality drinking water. Infectious diseases caused by water factor account for about 80% of infectious diseases in the world. Drinking water does not meet sanitary and hygienic requirements, poses a threat of mass diseases of the population, increased mortality (especially among children).

Availability of high quality drinking water in quantities that meet basic human needs is one of the conditions for improving human health and sustainable development of the state. Any non-compliance with the drinking water quality standard can lead to adverse consequences for the health and well-being of the population. In this regard, it is important to assess the impact of water on the human body, and especially the villagers. Since the water factor contributes to the emergence and complication of more than 80% of somatic diseases, such as atherosclerosis and other non-communicable diseases [25].

Considering that the overwhelming majority of scientific research over the last 20 years has focused on studying the hygienic state of drinking water supply in industrial cities, the need for such research in rural areas becomes even more argued.

**Purpose and objectives of the study.** The aim of the work is the scientific substantiation of sanitary and hygienic measures aimed at improving the safety and quality of drinking water of centralized and decentralized water supply sources in rural settlements of Dnepropetrovsk region, based on the ecological and hygienic assessment of quality indicators of tap water and additional treated drinking water.

In order to achieve the objective of the study, the following **objectives** are envisioned:

1. to assess the quality of water from the Karachunovskoye reservoir - a source of centralized water supply to the population of the western (Krivoy Rog zone) urbanization, according to the level of average annual indicators of salt composition, general sanitary, chemical, organoleptic and toxicological indicators of chemical composition of water for a long-term observation period (1965 - 2012) years.
2. to determine the level of morbidity among adult population - residents of rural taxa of Dnepropetrovsk region for 6 - year observation period (2008 - 2013) years.
3. To carry out a comparative assessment of quality indicators of pretreated water from different firms - producers, which is produced in Krivoy Rog urbanization zone, and tap drinking water in 1 taxon (Krivoy Rog district).

**Subject of the study:** indicators of drinking water quality; indicators of morbidity of rural population; oral exposure of rural population to chemical compounds with drinking water.

**Methods of research:** retrospective epidemiological study (to analyze morbidity among adult population of rural taxa of the region); sanitary-toxicological, physicochemical (to determine the indicators of water quality from water supply sources); sanitary - statistical (for mathematical processing of obtained quantitative indicators, methods of variation statistics).

Statistical processing of the results was performed on a personal computer using standard statistical packages STATISTICA 6.0 (license number 74017-640-0000106-57362). Excel package (license number 74017-640-0000106-57285) was used for primary preparation of tables and intermediate calculations. The following parameters were calculated: mean values (M), errors of mean (m), median (Me), 2575% confidence interval (CI).

## SECTION 1: MATERIALS AND METHODS OF RESEARCH

To solve the set tasks we have conducted complex ecological and hygienic studies of water quality from the Karachunovskoye reservoir - a source of centralized water supply for the population of the western (Krivoy Rog zone) urbanization; studied the quality indicators of pretreated drinking water produced by various firms; carried out a retrospective assessment of the health of the adult population of rural taxa of Dnepropetrovsk region. At realization of the program of the given work adequate to the purposes and tasks methods of research were used: retrospective epidemiological research; chemical-analytical (atomic absorption spectrophotometry); sanitary-chemical (photocolorimetry); sanitary-statistical methods (mathematical processing of the received quantitative indicators, methods of variation statistics). Generalized information about the stages, methods and volumes of research is given in (Table 1).

According to the territorial distribution, 22 administrative districts of Dnepropetrovsk region were classified into 6 types of taxa, according to the "Scheme of planning the territory of Dnepropetrovsk region" [26]. Classification of territorial taxa was carried out according to the indicators that take into account the development potential of individual taxa, namely: convenience of transport and geographical location, provision of rural population with drinking water of guaranteed quality and natural resource potential, the level of development of transport network, labor potential and the level of economic, social, environmental and urban development.

*Table 1*

**Stages, methods and scope of research**

| No. n/a | Research phase | Research Methods | Scope of research |
|---|---|---|---|
| 1.0 | Determination of water quality from the Karachunovskoye reservoir - a source of centralized water supply for the population of the western (Krivoy Rog zone) urbanization :[1] | | |
| 1.1. | Studies of the salt composition of water from the Karachunovskoye Reservoir based on the levels of mean annual values (1965-2012) years | Sanitary and chemical: determination of total hardness, dry residue, sulphates, chlorides by photo-lorimetric methods | 7296 |
| 1.2. | Assessment of organoleptic and general sanitary chemical indicators of water quality from the Karachunovskoye reservoir for (2008 -2012) years | Organoleptic: odor at 200 - 600C, taste and aftertaste, color, turbidity | 1000 |

| | | | |
|---|---|---|---|
| | | Sanitary and chemical: determination of pH, alkalinity, permanganate oxidizability, bi-chromate oxidizability, BOD, dissolved oxygen, total organic carbon by photo-colorimetric methods | 1750 |
| 1.3. | Determination of indicators of chemical composition of water from the Karachunovskoye reservoir (2008 -2012) years | Sanitary chemical: determination of ammonia nitrogen, nitrite, nitrate, Mo, As, Zn, cyanide, Ni, Pb, CaPO4, Mg, Na+ - K+, Fe, Cd, Cu, F, Cr, silicon | 5500 |
| | | Acid, polyphosphates, SPAS, petroleum products, phenol by photocolorimetric and atomic absorption spectrophotometry methods | |
| 2.0 | Study of quality indicators of pretreated drinking water used by the population of the western (Krivoy Rog zone) urbanization :[1] | | |
| 2.1. | Study of quality indicators of pre-treated drinking water produced by the manufacturer Mizrahin LLC (2012-2014) years of observation | Organoleptic: odor at 200 - 600C, taste and aftertaste, color, turbidity, precipitate | 1301 |
| | | Sanitary and chemical: determination of total hardness, dry residue, chlorides, sulfates, total iron, total alkalinity, Mg, Zn, Cu, Mn, pH, F, Al, Ag, Pb, Cd, Hg, ammonia nitrogen, nitrites, nitrates, oxidizability by photocolorimetric and atomic absorption spectrophotometry methods. | 2602 |

| 2.2. | Study of quality indicators of pre-treated drinking water produced by the producer company "Anisimov" LLC (2012-2014) years of observation | Organoleptic: odor at 200 - 600C, taste and aftertaste, color, turbidity, precipitate | 1059 |
|---|---|---|---|
| | | Sanitary chemical: determination of total hardness, dry residue, chlorides, sulfates, total iron, total alkalinity, Mg, Zn, Cu, Mn, pH, F, Al, Ag, Pb, Cd, Hg, ammonia nitrogen, nitrites, nitrates, oxidizability by photocolorimetric and atomic absorption spectro- photometry methods. | 2118 |
| 3.0 | Study of the dynamics of health indicators of the rural population of Dnipropetrovsk region for (2008 - 2013) years :[2] | | |
| 3.1. | Study of morbidity among adult population in 6 rural taxa of Dnepropetrovsk region, according to the levels of average multiyear indicators | Retrospective epidemiologic study: All diseases, I (A00- B99), II (C00-D48) III (D50-D89), (D50- D53), IV (E00-E90), VI (G00-G99), IX (I00-I99), X (J00- J99), XI (K00-K93), XII (L00-L99), XIII (M00-M99), XIV (N00-N99), XVII (Q00-Q99), XVII (Q20-Q28) classes of diseases (ICD - X). | 522720 |

**Classification of rural taxa of Dnepropetrovsk region.**

The first type - taxa with a high potential indicator and a high level of socio-economic and urban development (Krivoy Rog and Novomoskovsk districts); the second type - taxa with an average potential indicator and a high level of socio-economic and urban development (Nikopol and Pavlograd districts); the third type - taxa with a high potential indicator and an average level of socio-economic and urban development (Dnepropetrovsk district); the fourth type - taxa with an average potential indicator and an average level of socio-economic and urban development (Dnepropetrovsk district); the third type - taxa with a high potential indicator and an average level of socio-economic and urban development (Dnepropetrovsk district); the third type - taxa with a high potential indicator and an average level of socio-economic and urban development (Dnepropetrovsk district)

**The experimental zone - western (Krivoy Rog) urbanization zone** occupies (9% of the area of Dnepropetrovsk region, population - 740 thousand people, of which 94% - urban population). Krivoy Rog urbanization zone covers the city of Krivoy Rog, and the area of the Karachunovsky reservoir with water protection zones for the development of short-term and stationary recreation. The development of the city of Krivoy Rog and the area of the Karachunovskoye Reservoir is associated with the functioning of powerful mining and metallurgical industry enterprises, the indicators of urbanization and negative environmental impact have reached the crisis level. "Program of reforming and development of housing and communal services of Dnepropetrovsk region for 2004-2020" provides for: reconstruction of water supply and sewage networks; measures to introduce the latest technologies in the mining industry; reclamation of disturbed territories, landscaping and landscaping of SPZ; streamlining of transport and engineering communication network; determination of the area of water protection zones of the Karachunovskoye reservoir and their regime.

*Table 2*

**Structure of coverage of residents of rural taxa of Dnepropetrovsk oblast by centralized and decentralized drinking water supply**

| Rural taxon | Number of centralized sources of drinking water supply (abs., %) | Number of decentralized sources of drinking water supply (abs., %) | Total number of all sources of drinking water supply (abs., %) | Rank (by specific weight of coverage by both types of water supply sources) |
|---|---|---|---|---|
| 1 | 9 (4,8 %) | 235 (43,6 %) | 244 (33,6 %) | 1 |
| 2 | 13 (6,9 %) | 7 (1,3 %) | 20 (2,7 %) | 6 |
| 3 | 28 (15 %) | 5 (0,9 %) | 33 (4,5 %) | 5 |
| 4 | 42 (22,5 %) | 52 (9,7 %) | 94 (13 %) | 4 |
| 5 | 16 (8,5 %) | 91 (16,9 %) | 107 (14,7 %) | 3 |
| 6 | 79 (42,2 %) | 148 (27,5 %) | 227 (31,3 %) | 2 |
| Total by taxa | 187 (100 %) | 538 (100 %) | 725 (100 %) | |

In the study of drinking water quality indicators were used research methods: organoleptic - odor, color, turbidity; physico-chemical - total hardness, dry residue, chlorides, sulfates, total iron, copper, zinc, manganese, phenols, pH; sanitary-toxicological - nickel, arsenic, lead, fluorine, aluminum, selenium, mercury, nitrite nitrogen, nitrate nitrogen, acidity. In determining the organoleptic, physicochemical and sanitary-toxicological indicators we used the relevant normative documents (Table 3).

*Table 3*

**LIST OF DRINKING WATER QUALITY INDICATORS AND METHODS OF THEIR CONTROL**

| **Organoleptic indicators of drinking water quality** | |
|---|---|
| Odor at 20 °C | GOST 3351, DSTU EN 1420-1 |
| Odor when heated up to 60 °C | GOST 3351, DSTU EN 1420-1 |
| Taste and flavor | GOST 3351 |
| Colorfulness | GOST 3351, DSTU ISO 7887 |
| Turbidity | GOST 3351, DSTU ISO 7027 |
| **Chemical quality indicators affecting organoleptic properties drinking water properties** | |
| Inorganic components | |
| Hydrogen value (pH) | DSTU 4077 |
| Dry residue (total mineralization) | GOST 18164 |
| Total rigidity | GOST 4151, DSTU ISO 6059 |
| Total alkalinity | DSTU ISO 9963-1, DSTU ISO 9963-2 |
| Sulfates | GOST 4389, DSTU ISO 10304-1 |
| Chlorides | GOST 4245, DSTU ISO 10304-1, DSTU ISO 9297 |
| Total iron (Fe) | GOST 4011, DSTU ISO 6332 |
| Manganese (Mp) | GOST 4974, DSTU ISO 11885, DSTU ISO 15586 |
| Copper (C) | GOST 4388, DSTU ISO 11885, DSTU ISO 15586 |
| Zinc (Zn) | GOST 18293, DSTU ISO 11885, DSTU ISO 15586 |
| Calcium (Ca) | DSTU ISO 6058, DSTU ISO 11885 |
| Magnesium (Mg) | DSTU ISO 6059, DSTU ISO 11885 |
| Sodium (Na) | GOST 23268.6, DSTU ISO 11885 |
| Potassium (K) | GOST 23268.7, DSTU ISO 11885 |
| Organic components | |
| Petroleum products | GOST 17.1.4.01 |
| **Toxicological indicators of harmlessness of chemical composition drinking water** | |
| Inorganic components | |
| Aluminum (AH) | GOST 18165, DSTU KO 11885, DSTU ISO 15586 |
| Ammonia (NH4+) | GOST 4192, DSTU ISO 6778, DSTU ISO 7150-1, DSTU ISO 5664 |
| Cadmium (Cd) | DSTU ISO 11885, DSTU ISO 15586 |
| Arsenic (As) | GOST 4152, DSTU ISO 11885, DSTU ISO 15586 |
| Nickel (Ni) | DSTU 7150, DSTU ISO 11885 |
| Nitrates (NO3-) | GOST 18826, GOST 4192, DSTU 4078, DSTU ISO 7890-1, DSTU ISO 7890-2, |
| | DSTU ISO 10304-1 |
| Nitrites (NO2-) | GOST 4192, DSTU ISO 6777 |

| | |
|---|---|
| Mercury (Hg) | GOST 26927 |
| Lead(Pb) | GOST 18293, DSTU ISO 11885, DSTU ISO 15586 |
| Fluorides (F-) | GOST 4386, DSTU ISO 10304-1 |
| Total chromium (Cg) | DSTU ISO 11885, DSTU ISO 15586 |
| Cyanides (CN-) | DSTU ISO 6703-1, DSTU ISO 6703-2, DSTU ISO 6703-3 |
| Organic components | |
| Pesticides (total) | DSTU ISO 6468 |
| Synthetic surfactants (SPAS) | DSTU ISO 7875-1 |
| Integral indicators | |
| Permanganate oxidizability | GOST 23268.12 |
| Total organic carbon | DSTU EN 1484 |

In our study we used a set of sanitary-hygienic, epidemiological, physico-chemical, statistical methods. We determined average annual indicators of water quality from the surface water source - the Karachunovskoye reservoir, according to the requirements of SanPiN No. 4630-88 [27]. The water class of the water source for each of the studied indicators was determined according to GOST 4008:2007 [28]. The following were chosen as indicator indicators of water source water pollution: organoleptic (odor, taste and flavor, turbidity), total hardness, dry residue, sulfates, chlorides, permanganate oxidizability, pH, bichromate oxidizability, dissolved oxygen, total organic carbon, content of trace elements and chemical substances (Mo, As, Ni, Zn, Na+ - K+, Ca, Mg, Fe, Mn, Cu, F, cyanides, calcium phosphate, ammonia nitrogen, nitrites and nitrates, silicic acid, synthetic surfactants, polyphosphates and petroleum products) (a total of 33 indicators were studied). The study of most indicators of water quality from the Karachunovskoye reservoir was conducted during 2008-2012, salt composition of water (total hardness, dry residue, sulfates, manganese) according to the average annual values for the periods: 1965-1979, 1980-1990, 1991-2001, 2002-2012. The measurement of these indicators was carried out using gas chromatographic and atomic absorption methods.

During 2012-2014, we studied the quality of pretreated drinking water, which is produced by two specialized enterprises for pretreatment of water from the centralized water supply system in the city of Krivoy Rog - LLC "Mizrahin" and LLC "Anisimov". During the 3-year observation period, 3,903 tests of quality indicators of pretreated water from Mizrahin LLC and 3,177 tests of pretreated drinking water from Anisimov LLC were conducted. Pretreated drinking water produced by these specialized enterprises is used at local bottling points and provides water consumption for the population of Krivoy Rog city, and rural population of 1 taxon (Krivoy Rog rural district).

The average annual quality indicators of pretreated drinking water for 2012-2014 were compared with the current standards for packaged water from bottling points, according to GSanPiN 2.2.4-171-10 "Hygienic requirements for drinking water intended for human consumption" [29]. [29]. The quality of pretreated water was studied by organoleptic indicators: odor at $20^0$ and $60^0$ C, taste, color, turbidity, presence of sediment, physico-chemical indicators: total hardness, dry residue, total alkalinity, total iron, hydrogen index, sulfates, chlorides, sanitary-toxicological: copper, zinc, arsenic, manganese, lead, cadmium, aluminum, fluorides, oxidizability, ammonium, nitrites, nitrates (by NO ).[3]

Based on the data of official statistical reports [30], a database on the health status of the adult population living in 6 rural taxa of Dnepropetrovsk region was created.

The analysis of morbidity indicators among the adult population (according to 15 ICD-X classes) was conducted in 22 administrative districts of Dnipropetrovsk region, which were distributed into 6 types of rural taxa. The total number of resultant features (health indicators) that were analyzed are presented in (Table 1). The analysis was carried out by the method of retrospective continuous observation based on the reported data on the territory of 6 rural taxa of Dnipropetrovsk region, compared with the average annual indicators for Dnipropetrovsk region as a whole for the period 2008 - 2013. Statistical grouping of materials on the morbidity of rural population was carried out in accordance with the "International Statistical Classification of Diseases" (ICD-10) [31].

Statistical processing and analysis of the study results were carried out using methods of variation statistics [32] with the use of Microsoft Excel-2003 [33] and STATISTICA v. 6.1® (License No. 74017-640-0000106-57362). Statistical characteristics are presented as: number of observations (n), arithmetic mean (M), standard error of the mean (m), median (Me), relative indices (abs. number, %). Taking into account the law of data distribution (Kolmogorov-Smirnov test), Student's, Mann-Whitney, chi-square ($\chi2$), one-factor ANOVA and Kruskal-Wolis analysis of variance were used for comparison. The critical level of statistical significance (p) in testing statistical hypotheses was taken as ($p < 0.05$), ($p < 0.001$).

## SECTION 2: HYGIENIC ASSESSMENT OF DRINKING WATER QUALITY INDICATORS USED BY THE POPULATION OF THE WESTERN (KRIVOY ROG) URBANIZATION ZONE

There are more than 52.8 billion $m^3$ of water resources on the territory of Dnepropetrovsk region, including local runoff - 0.826 billion $m^3$ , groundwater reserves - 0.381 billion $m^3$ [34]. The main polluters of water bodies in the Dnieper River basin are industry (emissions in 2007 exceeded 790.9 million $m^3$ (62%), public utilities (359.5 million $m^3$ (28%), agriculture (123.4 million $m^3$ (9.6%), and other industries (1.6 million m3 (less than 1%) [35].

An important role in the accumulation of harmful substances in the Karachunovskoye reservoir is played by the inflow of polluted water into the Ingulets River from Kirovograd region, as heavy elements settle to the bottom at a sharp decrease in the water flow velocity in the reservoir in addition to the pollution that enters the river from the enterprises of the city of Krivoy Rog [36, 37]. The largest pollutants of water bodies in the Ingults basin upstream of the Karachunovskoye reservoir are effluents from industrial enterprises in Kirovograd and Dnepropetrovsk oblasts (Znamyanka, Alexandria, and Yellow Waters) and from agricultural enterprises [38].

The Krivoy Rog iron ore basin is the largest in Ukraine in terms of iron ore reserves and the main mining center of Dnepropetrovsk region. The city of Krivoy Rog concentrates 21 billion tons of iron ore reserves, of which industrial reserves amount to 18 billion tons [39]. The industrial and economic complex of the Krivoy Rog region was formed on the basis of the use of mineral resources, which influenced the development of production, led to a high territorial concentration of mining and metallurgical enterprises [40, 41]. [3] Annually operating mining enterprises in the basin pump out about 40 million m3 of groundwater (mine, open pit), among which 17-18 million m3 of highly mineralized mine water [42]. The maximum possibilities of groundwater use in the recycling cycles of mining enterprises border on the level of 28-29 million per year, the remaining 11-12 million $m^3$ annually temporarily accumulated and retained in the mine water reservoir [43].

The lack of a real alternative for the full use or utilization of excess recycled water dictates the need for annual use of measures for the discharge of excess recycled water from Kryvbas mining enterprises into the water bodies of the region [44].

A significant concentration of potentially hazardous objects on the territory of the Krivoy Rog region (mines, quarries, dumps, tailing ponds, slag heaps), provided that groundwater pumping is stopped or storage tanks overflow, will inevitably become a source of development of large-scale man-made disasters [45]. The infrastructure of the city of Krivoy Rog is associated with the functioning of powerful mining and metallurgical enterprises, which have reached a critical level in terms of urbanization and negative environmental impact [46].

**Dynamics of salt composition of water from the Karachunovskoye reservoir, by levels of mean annual values for (19652012) years**

The dynamics of total hardness growth in water from the Karachunovskoye reservoir according to the levels of average annual indicators was established: from (6.76±0.40) mmol/dm3 in 1965-1979 to (10.28±0.44) mmol/dm3 in 20022012. At the same time, during 1965-1979, according to GOST 4008:2007, water from the reservoir according to the indicator of total hardness belonged to the 3rd class of surface water supply sources, i.e. with "satisfactory, acceptable water quality" [28] [28]. According to the levels of average annual indicators during 1980-1990, 1991-2001, 2002-2012, the total hardness exceeded 7.0 mmol/dm$^3$ , i.e. water from the Karachunovskoye reservoir can be referred to the 4th class of surface water, i.e. "mediocre, limitedly suitable, undesirable water quality" (Fig. 1).

**Figure 1: Mean annual level of total hardness in the water of the Karachunovskoye reservoir, (mmol/dm$^3$ ).**

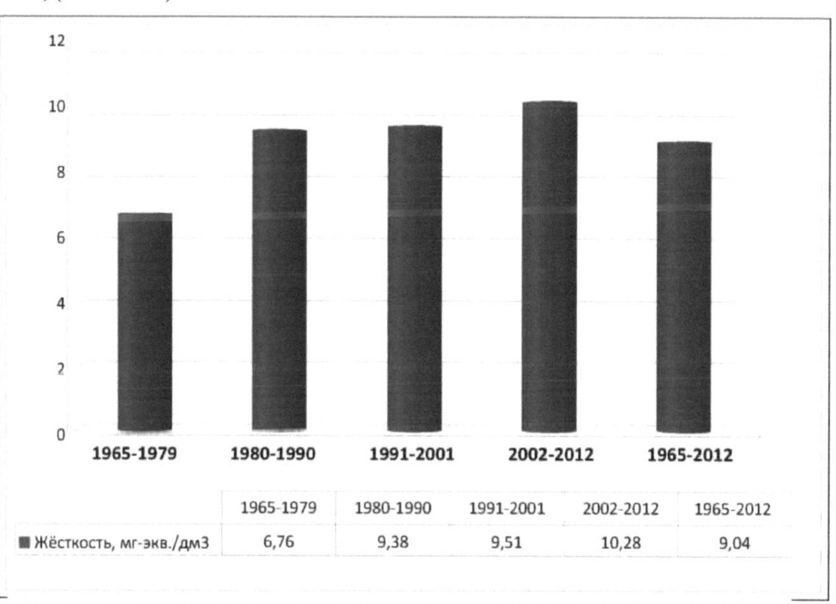


| | 1965-1979 | 1980-1990 | 1991-2001 | 2002-2012 | 1965-2012 |
|---|---|---|---|---|---|
| ■ Жёсткость, мг-экв./дм3 | 6,76 | 9,38 | 9,51 | 10,28 | 9,04 |

Dry residue for 1965-1979, 1980-1990 did not exceed the established hygienic standard (1000 mg/m$^3$ ) in accordance with SanPiN No. 4630-88 [27], and water from this reservoir was classified as class 3 according to GOST 4008:2007 [28]. From 1991 to 2012, the water quality in terms of dry residue content deteriorated, so the water source was categorized as a class 4 surface reservoir. For the similar period of observation, the dynamics of dry residue increase with exceeding the hygienic standard is shown: in 1991-2001, 1.04 times; in 2002-2012, 1.23 times. - in 1.23 times. The average content of dry residue for the period from 1965 to 2012 was at the level: 1005,31±37,12 mg/dm[3] (Fig. 2).

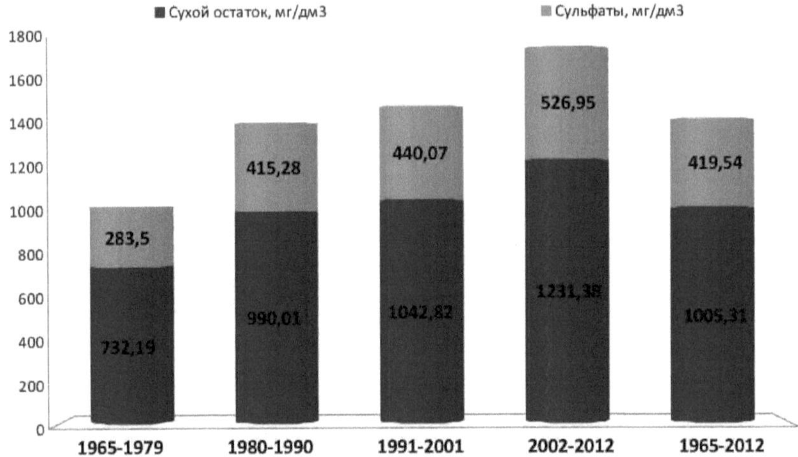


**Figure 2: Mean annual values of dry residue and sulfate in water from the Karachunovskoye Reservoir averaged over 19652012, (mg/dm ).[3]**

The tendency to increase the average annual indicator of sulfate content in water from the Karachunovskoye reservoir is shown. At that, the concentration of sulfates increased rapidly from 283.50±8.50 mg/dm[3] in 1965-1979 (exceeding MAC 1.13 times) to 526.95±6.27 mg/dm[3] in 20012012 (exceeding MAC 2.11 times). In terms of sulfate content, water from this reservoir belonged to class 4 of surface water bodies for the entire observation period (1965-2012). In terms of chloride content, a 1.34-fold decrease dynamics was noted: from 139.58±2.49 to 104.33±1.80 mg/dm[3] . During 2008-2012 chlorides did not exceed MAC (250 mg/dm[3] ) in the reservoir water, and water quality corresponded to class 3 (101-250 mg/dm[3] ). The highest manganese content was observed during 1980-1990 and 1991-2001 and ranged from 2.2-2.1 MAC. In general, the water quality from this water body belongs to class 3 and is at 0.162±0.018 mg/dm[3] for the entire observation period (1965-2012). The best quality of the surface water body in terms of manganese content (class 2) was recorded in 1965-1979 and 2001-2012 and was below the MAC level (0.1 mg/dm ).[3]

**Organoleptic and general sanitary chemical indicators of water quality from the Karachunovskoye reservoir for 2008 - 2012 years**

In terms of odor at 20-60°C, water belonged to class 1 in 2008-2012 (<1 point), except for 2009 (1 point), i.e. the reservoir water belonged to class 2. In general, the mean annual odor index of water from Karachunov reservoir belonged to class 1 of quality and was 0.77±0.05 points. Taste and flavor of water never exceeded hygienic norms and was within 0 points; water from this reservoir belonged to class 1 of surface water supply sources by quality.

The hydrogen index was within the established norm for class 2 surface sources (pH = 7.6-8.1) during the 5-year observation period, except for 2010 (pH = 8.21±0.06), when the water quality from the reservoir belonged to class 3 (pH = 8.2-8.5). An increasing trend in water color was found from 55.50±5.53 degrees in 2008 to 67.25±6.57 degrees in 2012, but the water from the reservoir belonged to class 2 of surface water quality (20-80 degrees) for the entire observation period.

The dynamics of increase of water turbidity in the reservoir was detected in 1.45 times - from 2.22±0.34 mg/dm$^3$ (2008) to 3.23±0.42 mg/dm$^3$ (2012), but by the level of this indicator the water was of the best quality, because it did not exceed the turbidity value for the 1st class of water supply sources (<20 mg/dm$^3$ ). The alkalinity indicator showed a decreasing trend during 2008-2012: from 4.50±0.05 to 4.19±0.06 mmol/dm$^3$ (1.07 times). In general, water from the Karachunovskoye reservoir according to this indicator belongs to the 3rd quality class (4.1-6.5 mmol/dm$^3$ ) for the whole period of observation.

Permanganate oxidizability ranged from 8.27±0.19 to 9.58±0.27 mg$O_2$ /dm3 with the highest value of the indicator in 2012 and a pronounced upward trend. However, in 2008-2012 the average annual permanganate oxidizability index was within the limits of class 2 (3-10 mgO /$dm_2^3$ ) and amounted to 8.65±0.11 mgO /$dm_2^3$ . In the water of the Karachunovskoye reservoir, there was a tendency for the bichromate oxidizability index (BOD) to decrease 1.38 times: from 21.72±0.67 mgO /$dm_2^3$ in 2008 to 15.75±0.79 mgO2/dm$^3$ in 2012. However, during the entire observation period, water quality from the reservoir was of class 2 (21.06±0.58 mgO /$dm_2^{3)}$ , not exceeding the established hygienic standard (9-30 mgC)2 dm3) (Fig. 3).

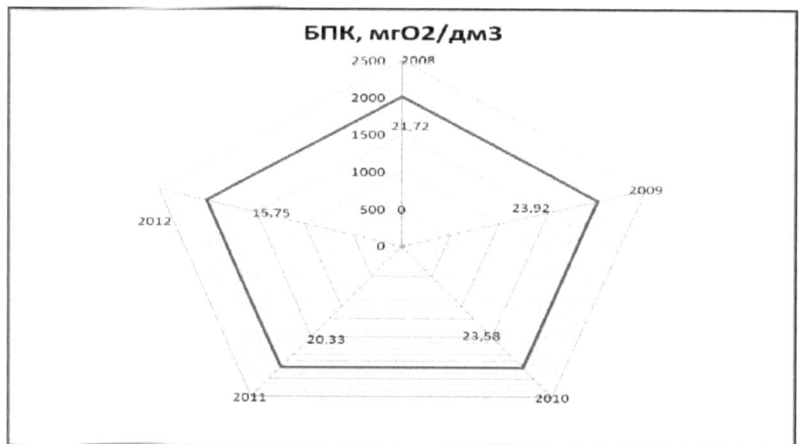


**Figure 3: Average value of bichromate oxidizability in the water of the Karachunovskoye**

**Reservoir during 2008-2012, (mgO /dm ).$_2^3$**

The BOD value showed an increasing trend in 2008-2012 with the highest level in 2011 - 2.81±0.35 mgO /dm$_2^3$ . At the same time, the average annual BOD indicator (2.58±0.18 mgO2/dm$^3$ ) did not exceed the fluctuation limits established for class 2 surface sources (1.3-3.0 mgO2/dm$^3$ ). Soluble oxygen in water from the reservoir did not exceed the limits of class 1 (>8.0 mgO2/dm$^3$ ), but over the 5-year observation period there was a tendency to increase its content in water - from 9.15±1.03 to 9.57±0.97 mgO$_2$ /dm3. According to the level of average annual indicator of soluble oxygen, water belongs to the 1st class of water sources quality (9.09±0.45 mgO2/dm$^3$ ) (Fig. 4).

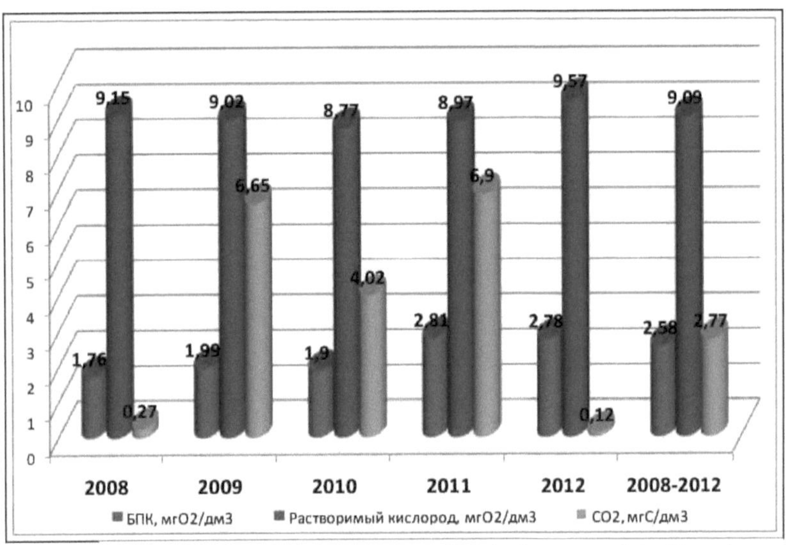


**Figure. 4. Average content of BOD, soluble oxygen, CO2 in the water of the Karachunovskoye reservoir during 2008-2012, (mgO2/dm ).$^3$**

The average content of total organic carbon in water was within 1-2 class, but according to the level of average annual indicator (2.77±0.63 mgC/dm$^3$ ) water from the Karachunovskoye reservoir was classified as quality class 1 (<5.0 mgC/dm$^3$ ). The highest value of total organic carbon was recorded in 2011 (6.90±0.96 mgC/dm$^3$ ; class 2), the lowest - in 2012 (0.12±0.08 mgC/dm$^3$ ; class 1).

**Toxicological indicators of the chemical composition of water from the Karachunovskoye Reservoir for 2008-2012**

The average molybdenum content in water did not exceed the MAC for surface water bodies (0.25 mg/dm$^3$ ), but water quality by this indicator belonged to class 3 for all years of observation except 2009 (<0.001 mg/dm$^3$ ), i.e. water in the reservoir corresponded to class 1 (<1 μg/dm$^3$ ). According to the level of annual average molybdenum (0.036±0.006) mg/dm$^3$ , the water was characterized as "satisfactory, acceptable quality" (class 3). Arsenic in the reservoir water did not exceed MAC (0.05 mg/dm$^3$ ) for 2008-2012, corresponding to water quality of class 2. A tendency to decrease the average arsenic content in water from the surface reservoir over the 5-year observation period was established, with values ranging from 0.005 to 0.001 mg/dm$^3$ . The content of cyanide in water remained constant, in the range of 0.02-0.05 mg/dm$^3$ , with the average annual indicator was at the level of 0.035±0.015 mg/dm$^3$ . Thus, the water content of cyanide was of quality class 3 (11-50 μg/dm$^3$ ) and did not exceed MAC (0.1 mg/dm$^3$ ) for the entire observation period.

As presented in (Fig. 5), the average nickel content in the reservoir water constantly fluctuated with a characteristic tendency to increase this chemical element 15 times: from 0.004±0.002 mg/dm$^3$ in 2009 to 0.060±0.004 mg/dm$^3$ in 2012. It should be noted that nickel concentration in water never exceeded the MAC (0.1 mg/dm$^3$ ). According to the average annual indicator of nickel content (0.043±0.007) mg/dm$^3$ the water is classified as quality class 2 (20-50 μg/dm$^3$ ). Lead did not exceed MAC (0.03 mg/dm$^3$ ) in water, and its content was constantly at the level <0.001 mg/dm$^3$ , so water from the surface water source was of the best quality (class 1).

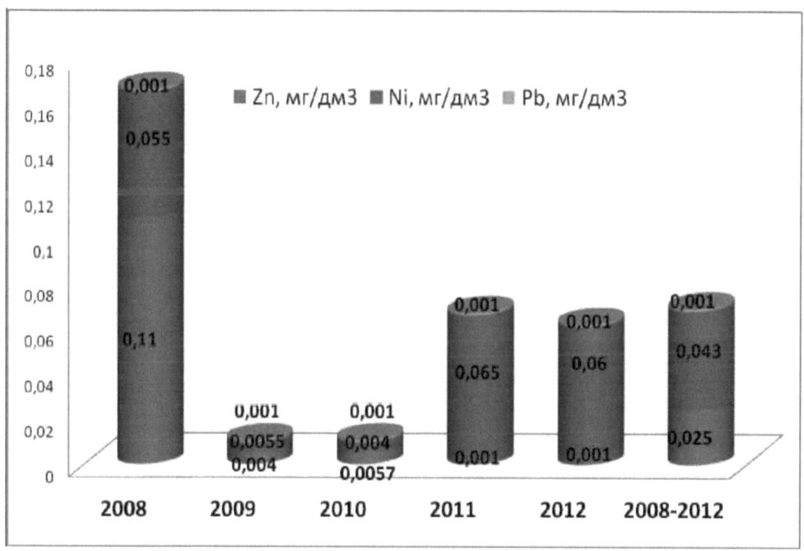


**Figure 5. Average content of heavy metals (Zn, Ni, Pb) in water from the Karachunovskoye reservoir for 2008-2012, (mg/dm$^3$ ).**

The average zinc content in water did not exceed MAC (1.0 mg/dm$^3$ ). Water from the Karachunovskoye Reservoir was characterized by "excellent, desirable water quality" (class 1) from 2009 to 2012, and satisfactory water quality (class 3) was found in 2008 at <0.11 mg/dm$^3$ . In terms of annual average zinc levels, water from the reservoir was predominantly characterized by "good, acceptable quality" (Class 2), within an average zinc concentration of 0.025±0.02 mg/dm .$^3$

Average calcium phosphate content exceeded MAC (3.5 mg/dm$^3$ ): 26.05 times (in 2008), and 23.5 times (in 2012). The average annual calcium phosphate amounted to 90.25±1.19 mg/dm$^3$ , exceeding MAC 25.78 times. The content of magnesium compounds in the reservoir water constantly exceeded the MAC for 2008-2012 and ranged from 76.57±1.19 to 58.85±2.64 mg/dm$^3$ (MAC 3.82-2.94 with a tendency to decrease in 2012). Magnesium compounds exceeded the hygienic standard (3.58 MAC) according to the level of the average annual indicator (71.59±1.36 mg/dm$^3$ ), so the water from the Karachunovskoye reservoir according to this indicator is classified as quality class 3.

The dynamics of decrease of sodium-potassium compounds in the reservoir water was shown: from 236.58±4.83 to 189.33±6.05 mg/dm$^3$ . However, the content of these compounds in water exceeded the MAC during the 5-year period and fluctuated within 1.18-1.11 MAC, except for 2011-2012. The mean annual sodium-potassium concentration in water also exceeded the MAC 1.07 times, amounting to 215.0±4.31 mg/dm$^3$ (Fig. 6).

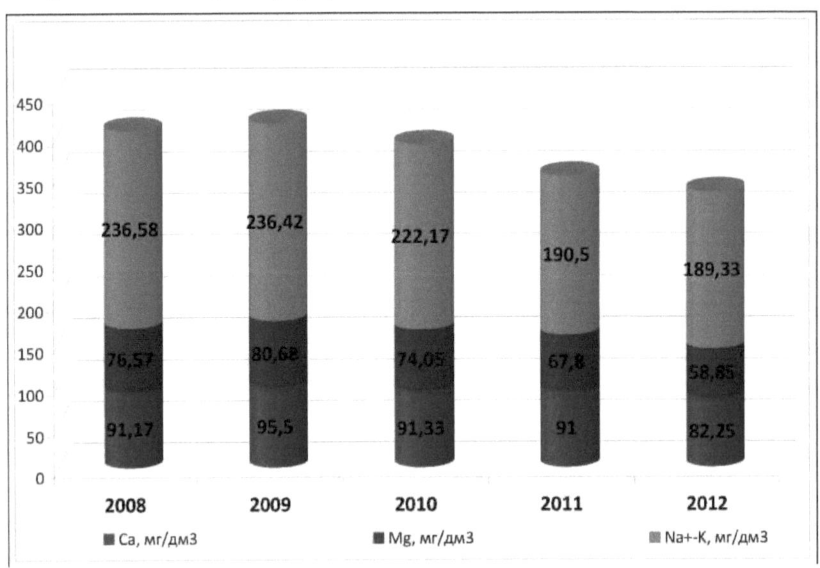


**Figure. 6. Average content of inorganic components in water from the Karachunovskoye reservoir for 2008-2012, (mg/dm ).³**

Ammonium nitrogen did not exceed the MAC value (2 mgCdm$^3$ ), but there was a tendency of increase in the content of this compound in 2008-2012 with the highest level in 2010 - 0.393±0.025 мгN/дм$^3$ . At the same time, water quality in 2010-2011 corresponded to class 3, while in previous years it corresponded to class 2. According to the level of the average annual indicator (within

0.262±0.013 мгN/дм$^3$ ) ammonium nitrogen corresponded to the 2nd class of water source quality 0.10-0.30 мгN/дм$^3$ . Nitrite nitrogen did not exceed the MAC value (3.3 мгN/дм$^3$ ) for the entire observation period, and the water was mainly of quality class 3. However, in 2008 and 2010, water from the Karachunovskoye reservoir corresponded to class 4 as "mediocre, limitedly suitable, undesirable quality" (>0.050 мгN/дм$^3$ ), with the highest value of this indicator in 2010 (0.061±0.021) мгN/дм$^3$ . It should be noted that the nitrate nitrogen content showed a negative downward trend in 2008-2012, but the concentrations of these compounds did not exceed the MAC value (45 мгN/дм$^3$ ). Water from the Karachunovskoye reservoir for the whole observation period can be classified as quality class 4 (>1.00 мгN/дм$^3$ ), with high nitrate nitrogen content in 2008 - 1.58±0.17 мгN/дм$^3$ (Table 4).

*Table 4* **Dynamics of indicators of nitrifying activity,** iron and copper content in the water of the Karachunovskoye reservoir for 2008-2012.

| Years | Ammonia nitrogen, мгN/дм$^3$ | Nitrite nitrogen, мгN/дм$^3$ | Nitrate nitrogen, мгN/дм$^3$ | Iron, mg/dm$^3$ | Copper, mg/dm$^3$ |
|---|---|---|---|---|---|
| **Mean value of the indicator, M±m** | | | | | |
| 2008 | 0,20±0,02 Me = 0.2 (25-75) % CI 0.125-0.275 | 0.058±0.030 Me = 0.02 (25-75) % CI 0.02-0.043 | 1.58±0.17 Me = 1.5 (25-75) % CI 1,175-1,9 | 0.026±0.003 Me = 0.02 (25-75) % CI 0.02-0.03 | 0.0056±0.001 Me = 0.005 (25-75) % DI 0.0025-0.0082 |
| 2009 | 0.22±0.02 Me = 0.22 (25-75) %DI 0.15-0.25 | 0.033±0.009 Me = 0.02 (25-75) % CI 0.02-0.031 | 1.23±0.16 Me = 1.15 (25-75) % CI 0.835-1.65 | 0.024±0.009 Me = 0.02 (25-75) % CI 0.02-0.03 | 0.0076±0.0026 Me = 0.005 (25-75) % CI 0.0025-0.0082 |
| 2010 | 0.208±0.023 Me = 0.185 (25-75) % CI 0.145-0.255 | 0.061±0.021 Me = 0.03 (25-75) % CI 0.02-0.0565 | 1.204±0.199 Me = 0.975 (25-75) % CI 0.59-1.8 | 0.342±0.003 Me = 0.035 (25-75) % CI 0.02-0.045 | 0.0025±0.0005 Me = 0.002 (25-75) % CI 0.001-0.004 |
| 2011 | 0.393±0.025 Me = 0.365 (25-75) % CI 0.335-0.43 | 0.033±0.010 Me = 0.02 (25-75) % CI 0.02-0.025 | 1.002±0.076 Me = 0.955 (25-75) % CI 0.8-1.14 | 0.060±0.009 Me = 0.055 (25-75) % CI 0.04-0.065 | 0.0027±0.0006 Me = 0.002 (25-75) % CI 0.001-0.004 |
| 2012 | 0.373±0.025 Me = 0.38 (25-75) %DI 0.31-0.425 | 0.030±0.006 Me = 0.02 (25-75) % CI 0.02-0.03 | 1.09±0.13 Me = 0.94 (25-75) % CI 0.735-1.365 | 0.083±0.021 Me = 0.055 (25-75) % CI 0.04-0.11 | 0.0031±0.0006 Me = 0.0025 (25-75) % CI 0.001-0.005 |
| **Annual averages for the 5-year period** | | | | | |
| 2008 - 2012 | 0.262±0.013 Me = 0.26 (25-75) %DI 0.18-0.32 | 0.043±0.008 Me = 0.02 (25-75) % CI 0.02 - 0.033 | 1.223±0.071 Me = 1.1 (25-75) % CI 0.81 - 1.55 | 0.045±0.005 Me = 0.03 (25-75) % CI 0.02 - 0.05 | 0.014±0.006 Me = 0.008 (25-75) % CI 0.005 - 0.0225 |

Notes. M - mean values, m - errors of the mean, Me - median **(Me), CI - 25-75% confidence interval.**

The tendency of increase of average iron content in the reservoir water in 2008-2012 with exceeding MAC (0.3 mg/dm$^3$ ) 1.14 times in 2010 (0.342±0.003 mg/dm$^3$ ) was established. There was also a change in the water class of the surface source: class 1 in 2008-2010 and class 2 in

20112012 , with iron content ranging from 0.060±0.009 to 0.083±0.021 mg/dm$^3$ . Cadmium in water was detected below MAC (<0.001 mg/dm$^3$ ) in all years of observation, with the water supply source corresponding to class 3 (0.6-5.0 μg/dm ).[3]

In the water of the Karachunovskoye reservoir during 2008-2012 there was a 1.8-fold decrease in copper content: from 0.0056±0.001 to 0.0031±0.0006 mg/dm$^3$ , but the compounds of this chemical element did not exceed the MAC value (1.0 mg/dm$^3$ ), and such water corresponded to class 2 (1-25 μg/dm$^3$ ) in terms of quality. Fluoride in the reservoir water did not exceed the MAC value (0.7 mg/dm$^3$ ), and the water corresponded to quality class 1 (<700 μg/dm$^3$ ). During the 5-year observation period, there was a 1.18-fold decrease in the content of fluorine compounds: from 0.313±0.021 to 0.266±0.164 mg/dm$^3$ , with the highest value in 2009. - 0.332±0.021 mg/dm$^3$ . Chromium content did not exceed the MAC (0.5 mg/dm$^3$ ) and was consistently <0.001 mg/dm$^3$ . According to the annual average of chromium compounds (0.030±0.006 mg/dm$^3$ ) the water belonged to class 1. A similar trend was observed for volatile phenols, which were below the MAC (<0.001 mg/dm$^3$ ) in 20082012 (quality class 1).

For the content of silicon compounds there was a pronounced tendency to its decrease from 2008 to 2012 from 6.175±1.414 to 5.725±1.519 mg/dm$^3$ . In some years there was observed an excess of hygienic standard of this chemical substance: in 2009 (1.14 MAC), in 2010 (1.27 MAC), in 2011 (1.05 MAC), with the highest value of silicic acid in 2010. - 12.683±0.751 mg/dm$^3$ . Polyphosphate content in water was well below the MAC (3.5 mg/dm$^3$ ), with a decreasing trend in 2008-2012. However, the highest level of polyphosphates was detected in 2008. - 0.53±0.05 mg/dm$^3$ , with a gradual decrease of these compounds from the beginning of 2011 - 0.14±0.03 mg/dm .[3]

SPAVs from 2008 to 2009 were at the level of (<0.001 mg/dm$^3$ ), water belonged to class 1 (<10 μg/dm$^3$ ). In the following years of observation, the water belonged to class 2 quality as the content of SPAV decreased 1.47 times: from 0.047±0.012 in 2011 to 0.032±0.009 mg/dm$^3$ in 2012. Petroleum products never exceeded the MAC value (0.3 mg/dm$^3$ ). During the 5-year observation period the dynamics of decrease in the content of these compounds in 1.2 times in the reservoir water was revealed: from 0.113±0.009 to 0.094±0.007 mg/dm$^3$ , with the highest value in 2012. Thus, the water from the Karachunovskoye reservoir by the content of oil products belongs to the 3rd quality class (51-200 μg/dm ).[3]

In the water of the Karachunovskoye reservoir over a long period of observation (from 1965 to 2012), an unfavorable trend of increasing salt composition, total hardness, dry residue, sulfates and chlorides was observed.) unfavorable tendency to increase salt composition, total hardness, dry residue content, sulfates and chlorides was noted, which is caused by systematic discharge of highly mineralized mine water from mining enterprises of Krivoy Rog city into the Ingulets and Saksagan rivers and subsequent pollution of the Karachunovskoye reservoir - the main water source of centralized domestic and drinking water supply for 94% of the urban population. In general, in terms of salt composition, water from the Karachunovskoye reservoir in some years of observation belonged to the 4th class of surface water quality as "mediocre, limited usable, undesirable quality".

A characteristic feature of Krivoy Rog urbanization zone is the presence of priority heavy metals (Mo, Mg, Cd, Ni, Zn, Fe, Cu, Pb, Cr) in water sources, which is due to intensive mining of iron ore. For example, the average iron content in 2010 was 0.342±0.003 $mg/dm^3$ , exceeding MAC (0.3 $mg/dm^3$ ) 1.14 times. The average manganese content exceeded the hygienic standard in 2008-2010 (MPC 1.42, 1.3 and 1.54, respectively), which is due to the high background content of this chemical element in the environmental objects of the industrial city and the annual discharge of highly mineralized mine water into local water sources.

## SECTION 3: MORBIDITY OF RURAL RESIDENTS IN SOME TAXA OF DNEPROPETROVSK OBLAST (BY LEVELS OF AVERAGE ANNUAL INDICATORS)

### Characterization of morbidity rate among adult population in separate taxa of Dnepropetrovsk region for (2008 - 2013) years

The highest specific weight of infectious and parasitic diseases was found among the adult population of taxa 1 (2.70 %) and 6 (2.60 %). As presented in (Fig. 7), the lowest incidence rate of class I diseases was reliably observed among the adult population of taxon 4: (72.98±6.05) ‰o ($p < 0.001$), with characteristic negative growth rates both by districts (-39.1 %) and by region (-75.0 %).

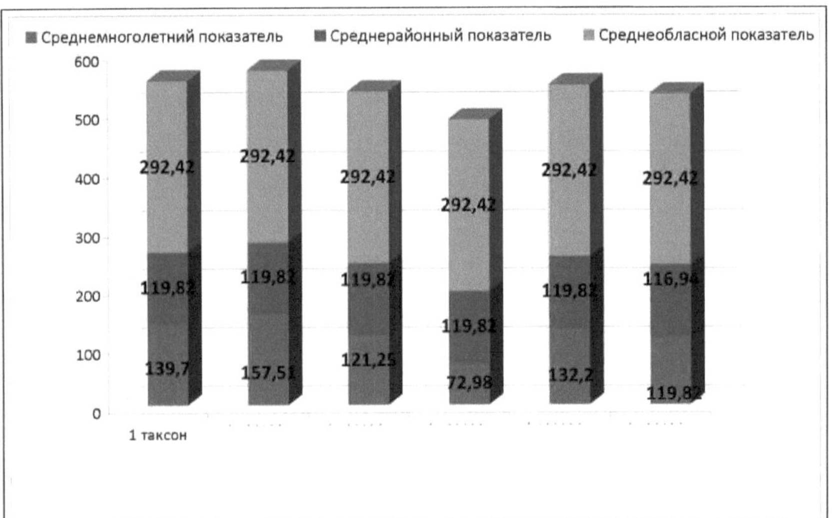


**Figure 7: Incidence of infectious and parasitic diseases in the adult population, according to the level of long-term averages, in individual taxa of Dnipropetrovsk region during 2008 -2013 (cases per 10,000 population).**

High intensity of class I diseases was reliably found among the rural population of taxon 2: (157.51±22.47) ‰o ($p < 0.001$), with an excess of the average regional morbidity rate by 1.31 times. The growth rate of infectious and parasitic diseases in the 2nd taxon by districts amounted to +31,4 %, by region -46,1 %. A similar trend was also found in the incidence rate of anemia among adult residents of individual taxa of Dnipropetrovsk Oblast (Fig. 8).

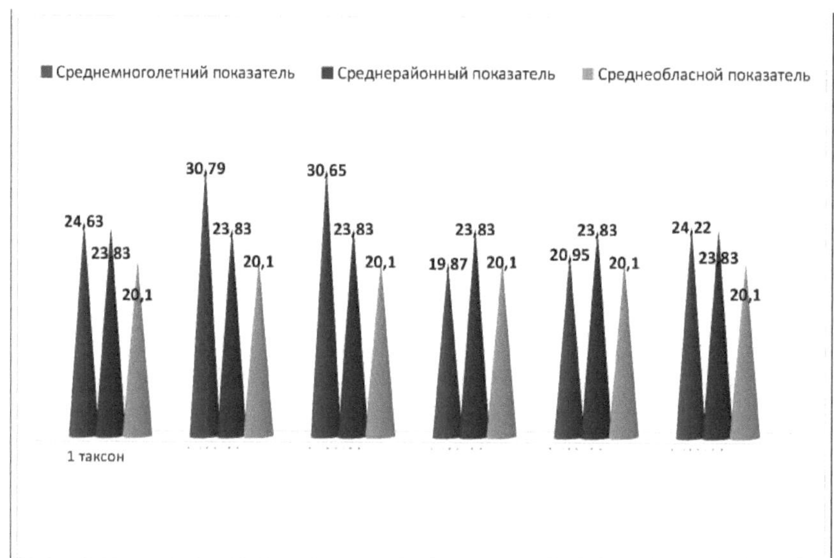


**Figure 8. Incidence of anemia in the adult population, according to the levels of long-term averages, in individual taxa of Dnipropetrovsk region during 2008 -2013 (cases per 10,000 population).**

The highest intensity of anemia was observed among rural residents of taxon 2: (30.79±5.62) ‰o , with the number of cases of incidence of class III diseases (D50-D53) being 1.29 times higher than the district average, and 1.53 times higher than the level of the regional average incidence rate. In taxon 2, positive growth rates of this class of diseases were registered both by districts (+29.2 %) and by region (+53.2 %). (Fig. 9) shows the growth rates of anemia morbidity among rural residents of individual taxa of Dnepropetrovsk region.

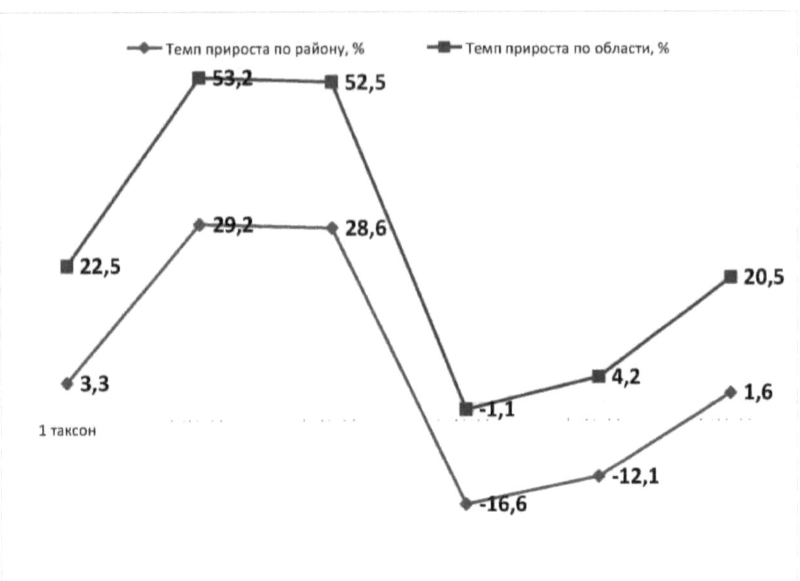


**Figure 9: Anemia growth rates among adults in selected taxa of Dnipropetrovsk region during 2008 - 2013.**

So, according to the rate of increase of class III diseases (D50-D53), the number of anemia cases increased rapidly among rural residents of taxa 1 to 3, with a characteristic tendency to decrease anemia among adults of taxa 4 and 5, and a characteristic positive rate of increase among residents of taxa 6, averaged over both districts and the region.

In the structure of all diseases the specific weight of cholelithiasis varies from 0.12 % in taxon 1 to 0.16 % in taxon 6. At that, the highest growth rates of XI class of diseases were observed in taxon 3 both by raions (+24,7 %) and by oblast (+0,8 %). The lowest incidence rate of cholelithiasis was reliably found among adult residents of taxon 1: (6.08±0.55) %00 ($p < 0.001$), with negative growth rates ranging from -21.2 to -36.3 % by districts and by region, respectively (Fig. 10).

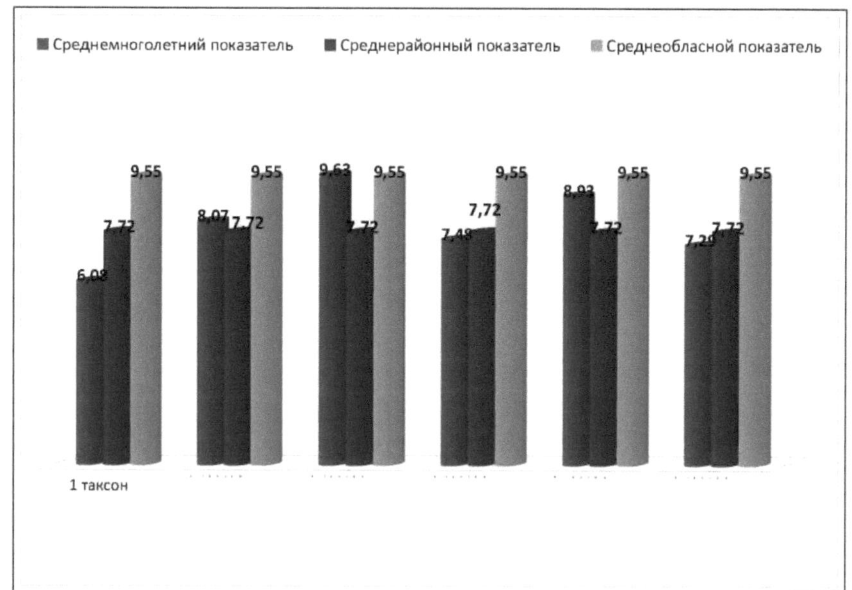


**Figure 10. Incidence of adult population with cholelithiasis, according to the levels of long-term averages, in separate taxa of Dnepropetrovsk region during 2008-2013 (cases per 10 000 population).**

The intensity of morbidity rates of XI class of diseases exceeded the level of average annual rates among rural residents of 2, 3, 5 taxa, respectively, by 1.04; 1.25 and 1.16 times. And only among residents of taxon 3 the level of morbidity of this class of diseases was significantly higher (9,63±0,54) ‱ ($p<0,05$) in comparison with the average regional indicator (9,55±0,30) ‱ by 1,0 times.

The incidence rate of salt arthropathy among the adult population was found to be higher in taxa 2, 3 and 4: (1.50 - 1.61) times; (2.95 - 3.17) times; (1.10 - 1.18) times than both district and oblast averages (Fig. 11). The highest among all types of taxa the rate of positive growth in XIV class of diseases (N25-N29) was observed among rural residents in taxon 3: +194.9 % (by districts), +216.8 % (by region).

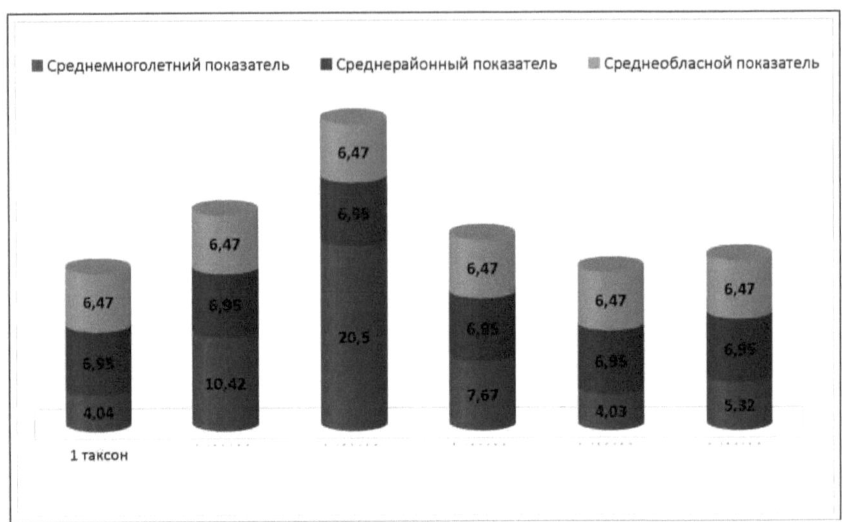


**Figure 11. Incidence of salt arthropathy in the adult population, according to the levels of long-term averages, in separate taxa of Dnepropetrovsk region during 2008-2013 (cases per 10 000 population).**

A completely different trend is observed in the level of morbidity of adult population with kidney and ureter stones, with the lowest level of intensity of XIV class of diseases (N17-N19) in 3 and 4 taxa: from (9,58±0,73) to (7,03±0,51) ‱ ($p < 0,001$). The highest level of morbidity of this class of diseases was determined among rural residents of taxon 2: (18,03±3,52)‱, with exceeding the average regional and average oblast indicators in 1,61 - 1,11 times (Fig. 12). At the same time, the rates of positive growth of kidney and ureter stones amounted to: +61,4 % in the districts and +10,9 in the region. The specific weight of XIV class of diseases (N17- N19) in separate taxa of the region amounted to: 0.23 % (in taxa 1 and 5); 0.31 % (in taxa 2); 0.16 % (in taxa 3 and 4); 0.26 % (in taxa 6).

Negative rates of increase in the incidence of kidney and ureteral stones were observed in taxa 3 and 4, both by district and by region: in the range from -14.2 to -41.0% in taxa 3; from -37.1 to -56.7% in taxa 4.

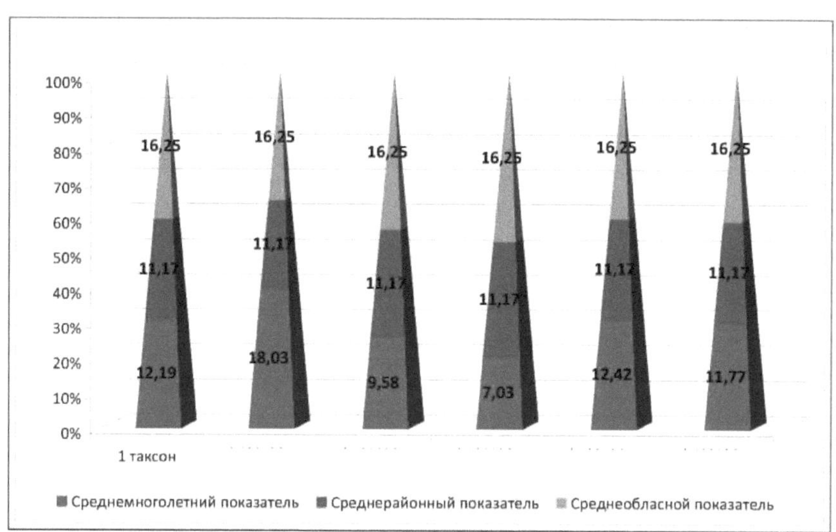


**Figure 12. Incidence of adult population with kidney and ureteral stones, according to the levels of long-term averages, in separate taxa of Dnepropetrovsk region during 2008-2013 (cases per 10 000 population).**

For diseases of skin and subcutaneous tissue, a trend of negative growth was revealed in all taxa by levels of average oblast indicators, while positive growth rates were observed in taxa 2, 3 and 5, on average by districts: by +20.4%; +74.4%; +13.3%, respectively (Fig. 13).

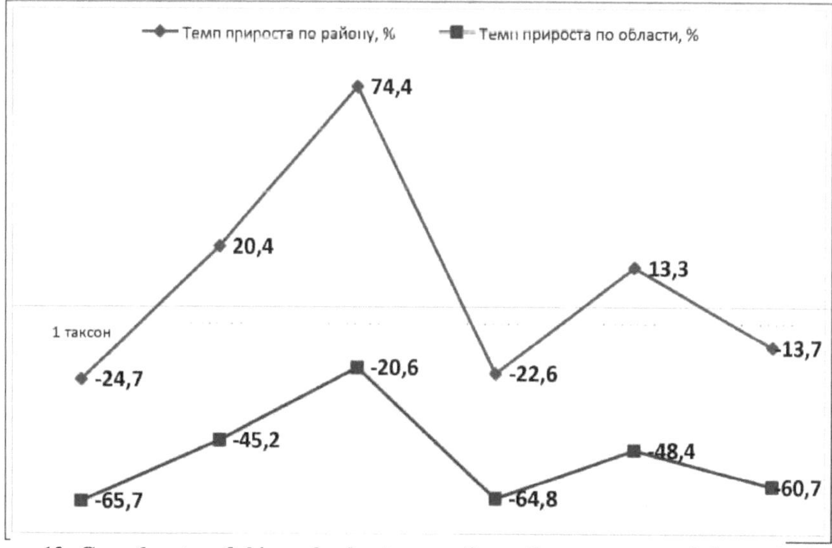


**Figure 13: Growth rates of skin and subcutaneous tissue diseases among adult population by**

**taxa in Dnipropetrovsk region during 2008 -2013.**

The highest level of morbidity of XII class of diseases was reliably detected among rural residents of taxon 3: (359,50±23,55) ‱ (p<0,05), with exceeding the average regional indicator 1,74 times. At the same time, the growth rates in 3 taxa amounted to: +74.4 % (by districts) and -20.6 % (by region). In the structure of all diseases the highest specific weight in this class of diseases is characteristic for taxon 3 (5,90 %), the lowest - for taxon 1 (3,00 %). The similar tendency was found also on intensive indices: the greatest number of cases of XII class of diseases was reliably observed among adult inhabitants of 3 taxon: (359,50±23,55) ‱ (p<0,05), the lowest - in 1 taxon: (155,30±26,71) ‱ (p<0,001).

Fig. 14 presents medical, demographic and economic losses associated with the negative impact of environmental factors. The specific weight of the water factor reaches 7% in the formation of economic losses: more than 450 billion hryvnias per year from adult morbidity; 18% causes the negative impact of the water factor on the morbidity of more than 6 million cases of diseases of different classes (circulatory, respiratory, digestive, blood and immune system, infectious diseases, etc.); 12% associated with the water factor causes 144 thousand deaths (due to diseases of the circulatory system, respiratory, neoplasms, etc.) [47 - 49]. [47 - 49].

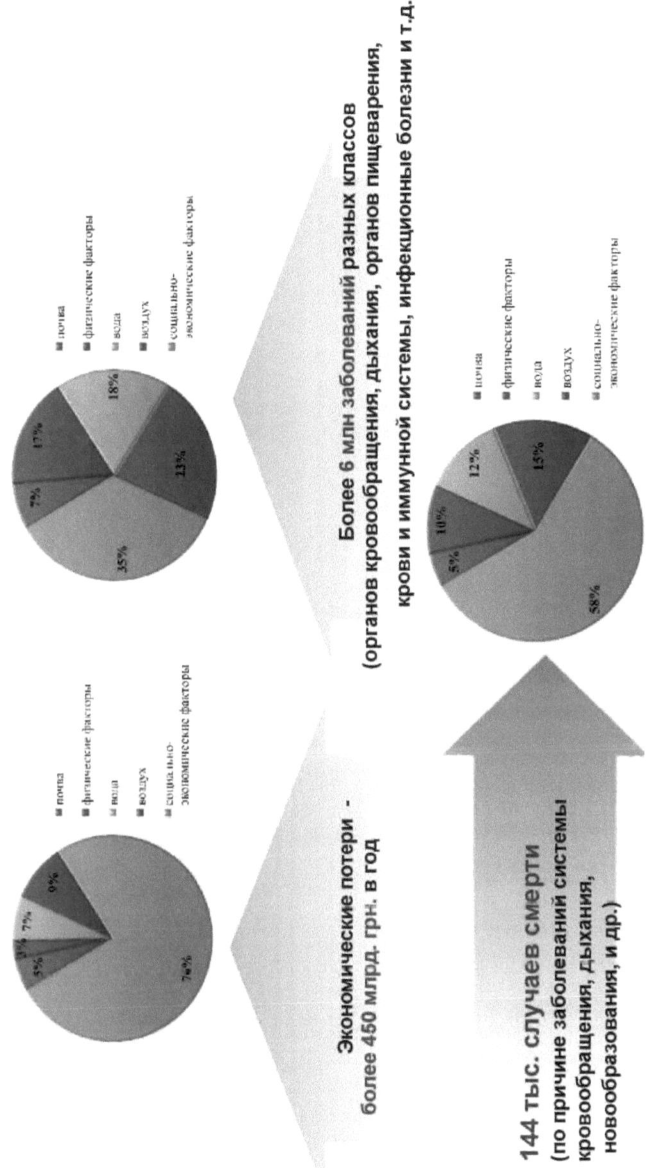


**Fig. 14: Medico-demographic and economic losses associated with the negative impact of environmental factors.**

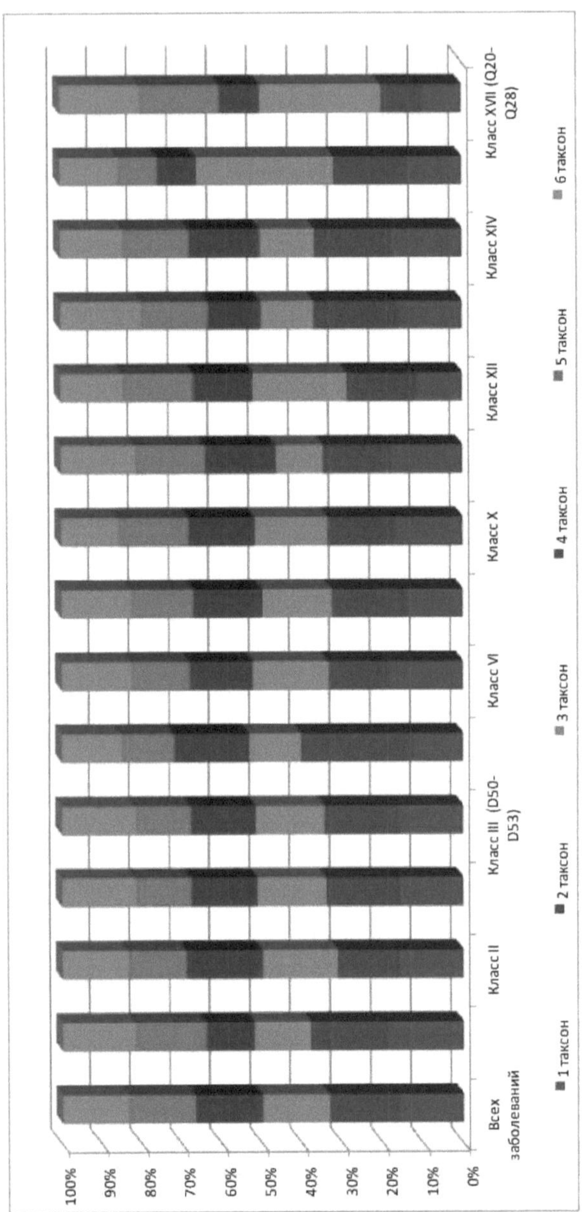


**Figure 15. Structure of adult morbidity by taxa of Dnepropetrovsk region, during 2008 -2013 (I, II, III, IV, VI, IX, IX, X, XI, XII, XIII, XIV, XVII) classes of ICD - X.**

## SECTION 4: COMPARATIVE CHARACTERISTIC OF QUALITY INDICATORS OF PRE-TREATED WATER OF DIFFERENT MANUFACTURERS, PRODUCED IN KRIVOY ROG URBANIZATION ZONE, AND TAP DRINKING WATER IN 1 RURAL TAXON (KRIVOY ROG DISTRICT)

Analyzed the quality of tap drinking water, which is consumed by the rural population of Krivoy Rog district (1 taxon), and pretreated water from different manufacturers (LLC "Mizrahin" and LLC "Anisimov"), produced in the Krivoy Rog urbanization zone, determined the efficiency of pretreatment in terms of total hardness, dry residue, chloride, sulfate, iron, pH, Cu, Zn, Mn, F, Al, ammonia nitrogen, nitrite and nitrate during 2012 - 2014. The results of our study indicate that the pretreatment of drinking water decreased the total hardness during the whole period of observation (Table 5).

*Table 5* **Comparative characterization of quality indicators of tap drinking water in 1 rural taxon (Krivoy Rog district) and drinking water pretreated from different firms - producers on indicators of total hardness, (mmol/dm )$^3$**

| Years | Pretreated drinking water Mizrahin LLC | Pretreated drinking water LLC Anisimov | Tap drinking water in 1 taxon | Efficiency of drinking water pretreatment by Mizrakhin LLC | Efficiency of additional treatment of drinking water of LLC "Anisimov" |
|---|---|---|---|---|---|
| 2012 | 2,31±0,11 | 3,17±0,31 | 1001,88±72,28 | 433,5 | 315,7 |
| 2013 | 2,23±0,02 | 2,38±0,27 | 5,72±0,70 | 2,56 | 2,40 |
| 2014 | 3,84±0,13 | 2,79±0,46 | 5,34±0,85 | 1,39 | 1,91 |
| p | p = 0,1991 | | | | |

Note.[1] p level of significance of efficiency of tap drinking water pretreatment from different firms - manufacturers by Pearson's criterion $\chi^2$ - Pearson.

Thus, the efficiency of water pretreatment for this indicator ranged from (433.5 to 315.7) MAC in 2012; from (2.56 to 2.40) MAC in 2013 and from (1.39 to 1.91) MAC in 2014, depending on the producer firm (p = 0.199). As presented in (Table 6), water pretreatment had a significant effect on drinking water quality as dry residue content decreased from 1.0 to 4.49 times for 2012 - 2014 (Mizrahin LLC pretreated water), and from 1.19 to 3.89 times (Anisimov LLC pretreated water). Thus, in the water pretreated from the first producer the content of dry residue decreased in dynamics by 1.2 times for the same period of observation: from (212.41±2.86) to (168.70±2.01) mg/dm$^3$ . In pretreated water from the second producer, this indicator increased 1.08 times: from (180.12±11.99) to (194.70±10.07) mg/dm .$^3$

*Table 6* **Comparative characterization of quality indicators of tap drinking water in 1 rural taxon (Krivoy Rog district) and drinking water pretreated from different firms - producers by dry residue, (mg/dm )[3]**

| Years | Pretreated drinking water Mizrahin LLC | Pretreated drinking water LLC Anisimov | Tap drinking water in 1 taxon | Efficiency of drinking water pretreatment by Mizrahin LLC | Efficiency of additional treatment of drinking water of Anisimov LLC |
|---|---|---|---|---|---|
| 2012 | 212,41±2,86 | 180,12±11,99 | 213,94±36,06 | 1,0 | 1,19 |
| 2013 | 214,50±2,23 | 210,70±3,27 | 619,71±99,95 | 2,89 | 2,94 |
| 2014 | 168,70±2,01 | 194,70±10,07 | 757,33±8,74 | 4,49 | 3,89 |
| p | p = 0,1991 | | | | |

Note.[1] p - level of significance of efficiency of tap drinking water pretreatment from different firms - manufacturers by Pearson's criterion $\chi^2$ - Pearson.

As can be seen (tab. 7), the efficiency of pretreatment of drinking water from the producer company "Mizrahin" LLC was significantly ($p < 0.05$) higher for chloride content: (26.4 MPC) in 2012, (10.5 MPC) in 2013, (2.85 MPC) in 2014, compared to pretreated water from the manufacturer "Anisimov" LLC: (9.74 MPC) in 2012, (5.53 MPC) in 2013, (6.21 MPC) in 2014.

*Table 7*

**Comparative characterization of quality indicators of tap drinking water in 1 rural taxon (Krivoy Rog district) and pre-treated drinking water from different manufacturers by chloride content, (mg/dm )[3]**

| Years | Pretreated drinking water Mizrahin LLC | Pretreated drinking water LLC Anisimov | Tap drinking water in 1 taxon | Efficiency of drinking water pretreatment by Mizrahin LLC | Efficiency of additional treatment of drinking water of Anisimov LLC |
|---|---|---|---|---|---|
| 2012 | 8,87±0,26 | 25,00±5,96 | 243,45±49,18 | 26,4 | 9,74 |
| 2013 | 8,49±0,18 | 16,20±3,30 | 89,59±16,25 | 10,5 | 5,53 |
| 2014 | 40,80±0,03 | 18,70±0,25 | 116,20±24,26 | 2,85 | 6,21 |
| p | p = 0.1991; p < 0.05[2] | | | | |

Note. 1p - level of significance of the efficiency of pretreatment **of tap drinking water from different firms - manufacturers by the $\chi^2$ - Pearson criterion;[2] - by one-factor ANOVA analysis of variance ($p < 0.05$).**

For sulfate content, the pretreatment efficiency varied within (3.04 - 2.03) MAC and from (1.24 to 2.81) MAC for 2012 - 2014, with the highest reduction of this indicator in 2013 (Table 8). Thus, sulfate content decreased by a factor of (9.9 to 10.5) after water pretreatment by both producer firms, as the highest content of this indicator was found in tap drinking water of 1 village taxon in 2013: (223.76±41.64) mg/dm$^3$ . At the same time, the content of sulfates fluctuated in drinking water after its additional treatment by different firms - producers, and never exceeded the MPC. In 2012, sulfates were recorded in the pretreated water of "Mizrahin" LLC at the concentration of (21.92±1.32) mg/dm$^3$ , while in 2014 at the level of (51.48±0.26) mg/dm$^3$ . A similar trend was found in the pretreated water of "Anisimov" LLC, with the highest value of this indicator in 2012: 53.68±12.54 mg/dm .³

*Table 8*

**Comparative characterization of quality indicators of tap drinking water in 1 rural taxon (Krivoy Rog district) and pre-treated drinking water from different manufacturers by** sulfate content, (mg/dm )³

| Years | Pretreated drinking water Mizrahin LLC | Pretreated drinking water OOO Anisimov | Tap drinking water in 1 taxon | Efficiency of drinking water pretreatment by Mizrahin LLC | Efficiency of additional treatment of drinking water of Anisimov LLC |
|---|---|---|---|---|---|
| 2012 | 21,92±1,32 | 53,68±12,54 | 66,65±2,22 | 3,04 | 1,24 |
| 2013 | 22,48±0,33 | 21,38±1,23 | 223,76±41,64 | 9,95 | 10,46 |
| 2014 | 51,48±0,26 | 37,18±1,37 | 104,37±3,50 | 2,03 | 2,81 |
| p | p = 0,1991 | | | | |

Note. 1p - significance level of aftertreatment efficiency tap drinking water from different manufacturers by Pearson's $\chi^2$ criterion.

In tap water from 1 taxon, sulfate content was the highest for all years of observation, compared to the quality of pre-treated drinking water. Pretreated water from 1 producer (Mizrahin LLC) had 3.04 times lower sulfate content than tap water in 2012; 10 times lower in 2013; and 2.02 times lower in 2014 than tap water from 1 taxon (p = 0.199). In the pretreated water from 2 producer (LLC "Anisimov"), sulfate content was 1.2 times lower than in tap water; in 2013 - 10.5 times lower; in 2014 - 3.0 times lower. The most effective was the additional treatment of tap drinking water of 1 taxon in terms of iron content in 2012 (Table 9).

*Table 9*

**Comparative characterization of quality indicators of tap drinking water in 1 rural taxon (Krivoy Rog district) and pre-treated drinking water from different manufacturers by iron content, (mg/dm )³**

| Years | Pretreated drinking water Mizrahin LLC | Pretreated drinking water LLC Anisimov | Tap drinking water in 1 taxon | Efficiency of drinking water pretreatment by Mizrahin LLC | Efficiency of additional treatment of drinking water of Anisimov LLC |
|---|---|---|---|---|---|
| 2012 | <0,2 | <0,2 | 0,027±0,011 | 7,4 | 7,4 |
| 2013 | <0,1 | <0,2 | <0,05 | 2 | 4 |
| 2014 | <0,1 | <0,1 | 0,06±0,01 | 1,6 | 1,6 |
| p | p = 1,0001 | | | | |

Note. 1p - level of significance of the efficiency of **tap drinking water** pretreatment **from different firms - manufacturers according to the criterion $\chi^2$ - Pearson.**

Efficiency of drinking water pretreatment by this indicator increased 7.4 times in 2012, 2.0 times in 2013 and 1.6 times in 2014. At the same time, the highest iron content in tap water was

0.06±0.01 mg/dm$^3$ in 2014, while the pretreated water was much lower than 0.1 mg/dm .$^3$

In 2012, the pH value was (1.08 - 1.02) times lower in the samples of pretreated water from both manufacturers than in tap drinking water: 7.70±0.06, while the pretreatment efficiency increased by (1.09 - 1.02) times. During 2013 - 2014, in the pretreated water from the producer "Mizrahin" LLC the pH value fluctuated within (1.07 - 1.05); while in the pretreated water from the second producer - "Anisimov" LLC the pH decreased in (1.05 - 1.04) times. As can be seen in (Table 10), the highest pH value in tap water was recorded in 2012 and amounted to 7.70±0.06, while the lowest value was recorded in 2014: 7.24±0.05 (p = 0.223).

*Table 10*

**Comparative characterization of quality indicators of tap drinking water in 1 rural taxon (Krivoy Rog district) and drinking water pretreated from different manufacturers by pH value**

| Years | Pretreated drinking water Mizrahin LLC | Pretreated drinking water LLC Anisimov | Tap drinking water in 1 taxon | Efficiency of drinking water pretreatment by Mizrahin LLC | Efficiency of additional treatment of drinking water of Anisimov LLC |
|---|---|---|---|---|---|
| 2012 | 7,09±0,02 | 7,52±0,14 | 7,70±0,06 | 1,09 | 1,02 |
| 2013 | 7,12±0,16 | 7,05±0,17 | 7,66±0,04 | 1,07 | 1,09 |
| 2014 | 7,59±0,07 | 7,52±0,12 | 7,24±0,05 | 1,05 | 1,04 |
| p | p = 0,2231 | | | | |

Note. 1p - level of significance of the efficiency of **tap drinking water** pretreatment **from different firms - manufacturers according to the criterion $\chi^2$ - Pearson.**

The results of our study indicate an improvement in the quality of pretreated drinking water in terms of TM (Cu, Zn, Mn) content, as presented in (Tables 11 - 13). Thus, after pretreatment of drinking water by the company - manufacturer LLC "Mizrahin" during 2012 - 2014, the content of copper decreased from (3.65 to 4.4) times, zinc decreased from (15.3 to 1.5) times, manganese fluctuated within (12.5 - 13) times. Efficiency of water pretreatment from the second producer - LLC "Anisimov" on the content of these TMs also increased: Cu - in (1.38 - 1.68) times, Zn - in (7.14 - 2.2) times, Mn - in (1.85 - 2.08) times.

*Table 11*

**Comparative characterization of quality indicators of tap drinking water in 1 rural taxon (Krivoy Rog district) and pre-treated drinking water from different firms - producers on copper content, (mg/dm )$^3$**

| Years | Pretreated drinking water Mizrahin LLC | Pretreated drinking water OOO Anisimov | Tap drinking water in 1 taxon | Efficiency of drinking water pretreatment by Mizrahin LLC | Efficiency of additional treatment of drinking water of Anisimov LLC |
|---|---|---|---|---|---|
| 2012 | 0,11±0,03 | 0,040±0,012 | 0,029±0,016 | 3,65 | 1,38 |

| 2013 | 0,0994±0,0006 | 0,085±0,009 | 0,016±0,008 | 6,19 | 5,31 |
|---|---|---|---|---|---|
| 2014 | 0,0053±0,0046 | 0,037±0,001 | 0,022±0,002 | 4,4 | 1,68 |
| p | p = 0,1991 | | | | |

Note. [1]p - significance level of aftertreatment efficiency
tap drinking water from different manufacturers by Pearson's $\chi^2$ criterion.

*Table 12*

**Comparative characterization of quality indicators of tap drinking water in 1 rural taxon (Krivoy Rog district) and pre-treated drinking water from different manufacturers by zinc content, (mg/dm )$^3$**

| Years | Pretreated drinking water Mizrahin LLC | Pretreated drinking water LLC Anisimov | Tap drinking water in 1 taxon | Efficiency of drinking water pretreatment by Mizrahin LLC | Efficiency of additional treatment of drinking water of Anisimov LLC |
|---|---|---|---|---|---|
| 2012 | 0,15±0,01 | 0,0014±0,0088 | <0,01 | 15,3 | 7,14 |
| 2013 | 0,0278±0,0069 | 0,046±0,012 | 0,024±0,003 | 1,16 | 1,92 |
| 2014 | 0,015±0,001 | 0,0045±0,0007 | <0,01 | 1,5 | 2,2 |
| p | p = 0,1991 | | | | |

Note.[1] p - significance level of aftertreatment efficiency
tap drinking water from different manufacturers by Pearson's $\chi^2$ criterion.

*Table 13*

**Comparative characterization of quality indicators of tap drinking water in 1 rural taxon (Krivoy Rog district) and pre-treated drinking water from different manufacturers by manganese content, (mg/dm )$^3$**

| Years | Pretreated drinking water Mizrahin LLC | Pretreated drinking water LLC Anisimov | Tap drinking water in 1 taxon | Efficiency of drinking water pretreatment by Mizrahin LLC | Efficiency of additional treatment of drinking water of Anisimov LLC |
|---|---|---|---|---|---|
| 2012 | <0,05 | 0,027±0,010 | <0,05 | 0 | 1,85 |
| 2013 | 0,004±0,003 | 0,054±0,027 | <0,05 | 12,5 | 1,08 |
| 2014 | 0,0043±0,0005 | 0,025±0,005 | 0,052±0,002 | 13 | 2,08 |
| p | p = 0,1991 | | | | |

Note. [1]p - significance level of aftertreatment efficiency
tap drinking water from different manufacturers by Pearson's $\chi^2$ criterion.

Of particular note is the fact that the content of TM (Cu, Zn, Mn) in the pretreated drinking water of both producer firms was significantly lower than in tap water of 1 taxon (Figs. 16, 17, 18).

In 2014, the manganese content was (12 - 2.08) times lower in pretreated water than in tap drinking water, while the pretreatment efficiency increased by (13 - 2.08) times. A similar trend was determined for copper content in 2014. This TM in pretreated water from the company - manufacturer LLC "Mizrahin" was 4.1 times lower than in tap water.

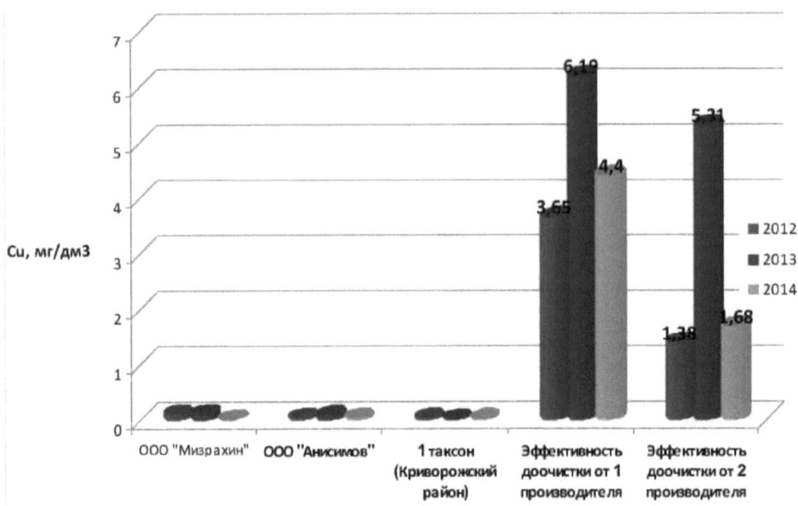


**Figure 16. Comparative characterization of tap water quality in Krivoy Rog district and pretreated drinking water from different manufacturers in terms of copper content.**

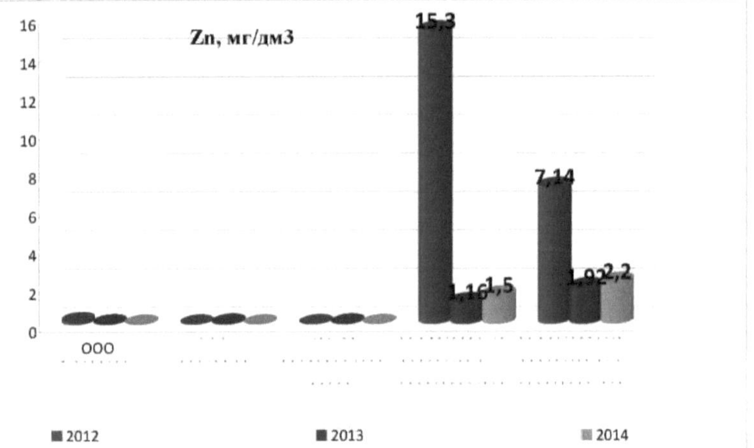


**Figure 17. Comparative characterization of tap water quality in Krivoy Rog district and pre-treated drinking water from different manufacturers in terms of zinc content.**

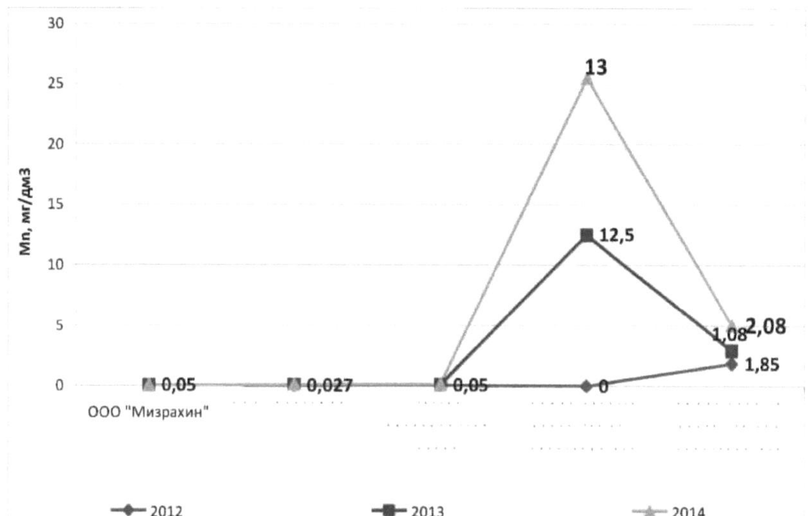


**Figure 18. Comparative characterization of tap water quality in Krivoy Rog district and pre-treated drinking water from different manufacturers in terms of manganese content.**

Regarding fluoride, pretreatment efficiency increased by (1.33 - 8.62) times in drinking water from 1 producer (Mizrahin LLC), by (1.22 - 8.62) times in water from 2 producers (Anisimov LLC) (Table 14).

*Table 14*

**Comparative characterization of quality indicators of tap drinking water in 1 rural taxon and pre-treated water from different manufacturers in terms of fluoride content (mg/dm )[3]**

| Years | Pretreated drinking water Mizrahin LLC | Pretreated drinking water LLC Anisimov | Tap drinking water in 1 taxon | Efficiency of drinking water pretreatment by Mizrahin LLC | Efficiency of additional treatment of drinking water of Anisimov LLC |
|---|---|---|---|---|---|
| 2012 | 0,13±0,06 | <0,08 | 0,098±0,018 | 1,33 | 1,22 |
| 2013 | <0,08 | <0,08 | 0,20±0,13 | 2,5 | 2,5 |
| 2014 | <0,08 | <0,08 | 0,69±0,01 | 8,62 | 8,62 |
| p | p = 0,1991 | | | | |

Note.[1] p - significance level of aftertreatment efficiency tap drinking water from different manufacturers by Pearson's $\chi^2$ criterion.

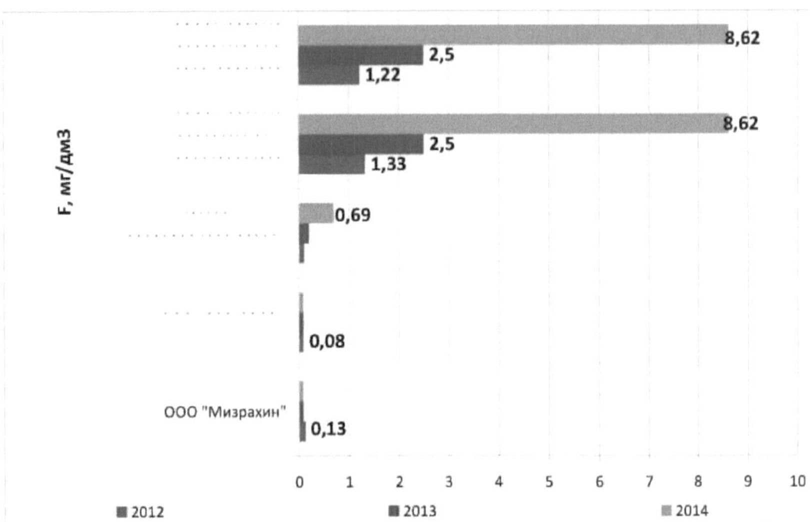


**Figure 19. Comparative characterization of tap water quality in Krivoy Rog district and pretreated drinking water from different manufacturers in terms of fluoride content.**

As presented in (Fig. 19), the highest fluoride content was found in tap drinking water of 1 taxon in 2014: 0.69±0.01 mg/dm$^3$ , while in water pretreated by both producers in some years of observation this indicator was at the level < 0.08 mg/dm$^3$ . The low aluminum content < 0.04 mg/dm$^3$ for all years of observation in water pretreated by both manufacturers is noteworthy (Table 15).

*Table 15*

**Comparative characteristics of quality indicators of tap drinking water in 1 rural taxon (Krivoy Rog district) and pre-treated drinking water from different manufacturers by aluminum content, (mg/dm )$^3$**

| Years | Pretreated drinking water Mizrahin LLC | Pretreated drinking water LLC Anisimov | Tap drinking water in 1 taxon | Efficiency of drinking water pretreatment by Mizrahin LLC | Efficiency of additional treatment of drinking water of Anisimov LLC |
|---|---|---|---|---|---|
| 2012 | < 0,04 | < 0,04 | < 0,05 | 1,25 | 1,25 |
| 2013 | < 0,04 | < 0,04 | 0,20±0,09 | 5,0 | 5,0 |
| 2014 | < 0,04 | < 0,04 | 0,13±0,05 | 3,25 | 3,25 |
| p | p = 0,1991 | | | | |

Note. 1p - significance level of aftertreatment efficiency tap drinking water from different manufacturers by Pearson's $\chi^2$ criterion.

In general, pre-treated drinking water, as well as tap water does not meet the requirements of GOST 7525:2014 [50], since aluminum should be absent in water of decentralized drinking water supply (non-packaged and packaged). Traces of the presence of this indicator were detected in both pretreated and tap water. Thus, in drinking water of 1 taxon in separate years of observation aluminum was within the limits: from (0,20±0,09) to (0,13±0,05) mg/dm$^3$ , decreased in dynamics in 1,5 times. At the same time, the efficiency of additional treatment of drinking water from both manufacturers increased in (1.25 - 3.25) times. The highest aluminum content was detected in tap water in 2013, while the pretreatment efficiency increased by 5.0 times (Fig. 20).

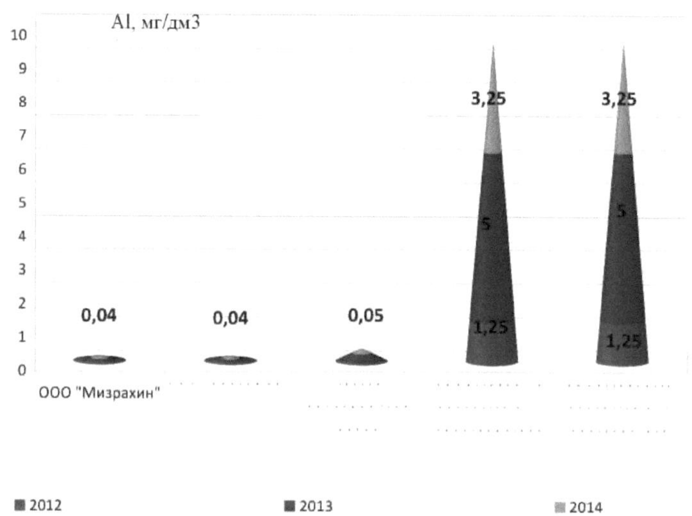


**Figure 20. Comparative characteristics of tap water quality in Krivoy Rog district and pretreated drinking water from different manufacturers by aluminum content.**

Some indices of nitrifying activity did not respond to requirements of the normative document GOST 7525:2014 [50] (Tables 16 - 18).

*Table 16*

**Comparative characterization of quality indicators of tap drinking water in 1 rural taxon (Krivoy Rog district) and pre-treated drinking water from different manufacturers by ammonia nitrogen content, (mg/dm )³**

| Years | Pretreated drinking water Mizrahin LLC | Pretreated drinking water LLC Anisimov | Tap drinking water in 1 taxon | Efficiency of drinking water pretreatment by Mizrahin LLC | Efficiency of additional treatment of drinking water of Anisimov LLC |
|---|---|---|---|---|---|
| 2012 | <0,05 | <0,05 | 0,019±0,011 | 2,6 | 2,6 |
| 2013 | <0,1 | <0,1 | 0,22±0,06 | 2,2 | 2,2 |
| 2014 | <0,1 | <0,1 | 0,31±0,05 | 3,1 | 3,1 |
| p | p = 0.223[1] ; p <0.001[2] | | | | |

Note.[1] p - level of significance of the efficiency of pretreatment of tap drinking water from different firms - manufacturers by the $\chi^2$ - Pearson criterion;[2] - by one-factor ANOVA analysis of variance (p < 0.001).

*Table 17*

**Comparative characterization of quality indicators of tap drinking water in 1 rural taxon**

**(Krivoy Rog district) and pre-treated drinking water from different manufacturers by nitrite content, (mg/dm )[3]**

| Years | Pretreated drinking water Mizrahin LLC | Pretreated drinking water LLC Anisimov | Tap drinking water in 1 taxon | Efficiency of drinking water pretreatment by Mizrahin LLC | Efficiency of additional treatment of drinking water of Anisimov LLC |
|---|---|---|---|---|---|
| 2012 | <0,02 | <0,02 | 15,45±0,04 | 772,5 | 772,5 |
| 2013 | 0,0 | 0,0 | 0,011±0,006 | - | - |
| 2014 | 0,0 | 0,0 | 0,031±0,014 | - | - |
| p | p = 0,2231 | | | | |

Note. 1p - level of significance of aftertreatment efficiency tap drinking water from different manufacturers by Pearson's $\chi^2$ criterion.

*Table 18*

**Comparative characterization of quality indicators of tap drinking water in 1 rural taxon (Krivoy Rog district) and pre-treated drinking water from different manufacturers by** nitrate content, (mg/dm )[3]

| Years | Pretreated drinking water Mizrahin LLC | Pretreated drinking water OOO Anisimov | Tap drinking water in 1 taxon | Efficiency of drinking water pretreatment by Mizrahin LLC | Efficiency of additional treatment of drinking water of Anisimov LLC |
|---|---|---|---|---|---|
| 2012 | <0,5 | <0,5 | 1,71±0,18 | 3,42 | 3,42 |
| 2013 | <0,5 | <0,5 | <0,5 | 1,0 | 1,0 |
| 2014 | <0,5 | <0,5 | 1,07±0,39 | 2,14 | 2,14 |
| p | p = 0,1991 | | | | |

Note.[1] p - significance level of aftertreatment efficiency tap drinking water from different manufacturers by Pearson's $\chi^2$ criterion.

Ammonia nitrogen was consistently detected in water pretreated from both manufacturers in concentrations of (<0.05 - 0.1) mg/dm$^3$ , as well as in tap drinking water in the range from (0.019±0.011) to (0.31±0.05) mg/dm$^3$ , with an increasing trend of 16.3 times during 2012 - 2014. At the same time it was shown a reliable efficiency of drinking water pretreatment from both firms - manufacturers in 2.6 - 3.1 times ($p < 0.001$) (Fig. 21).

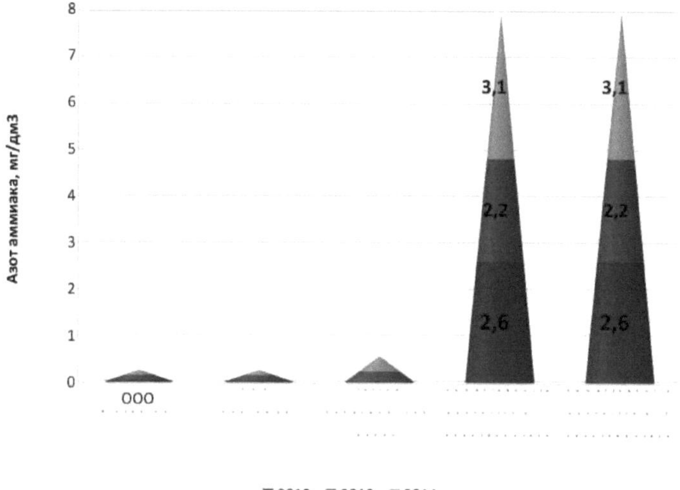


**Figure 21. Comparative characterization of tap water quality in Krivoy Rog district and pretreated drinking water from different manufacturers by nitrogen content ammonia.**

Nitrites exceeded the MAC in tap water of 1 taxon 772.5 times in 2012 and 1.5 times in 2014. In water pretreated from both producers, nitrite was within the MAC (< 0.02 mg/dm$^3$ ) in 2012, and was absent in 2013 - 2014 (p = 0.223) (Fig. 22).

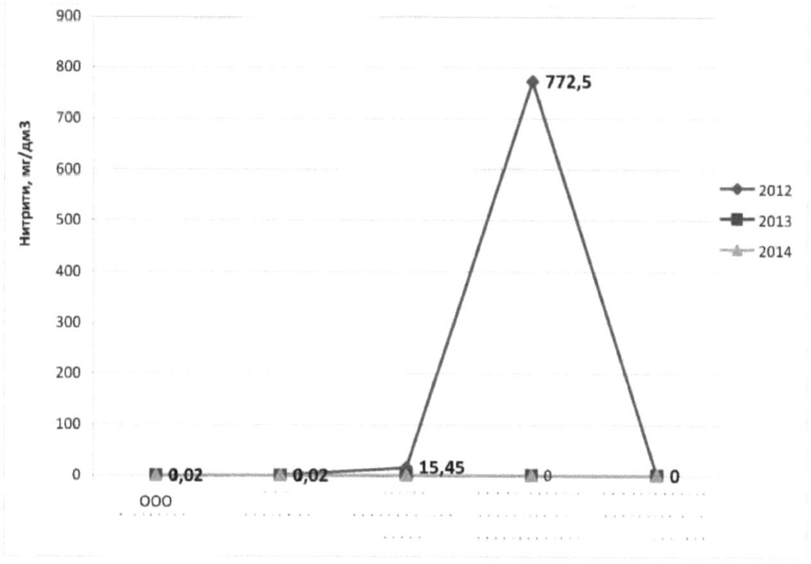

**Figure 22. Comparative characterization of tap water quality in Krivoy Rog district and pretreated drinking water from different manufacturers in terms of nitrite content.**

Nitrate content in pretreated and tap water did not exceed MAC during 2012 - 2014. Low concentrations of nitrates (< 0.5 mg/dm$^3$ ) were found in pretreated water, while in tap water nitrates were in the range from (1.71±0.18) to (1.07±0.39) mg/dm$^3$ , with a tendency to decrease by 1.6 times. Water pretreatment efficiency for this indicator increased from 3.42 times in 2012 to 2.14 times in 2014 (Fig. 23).

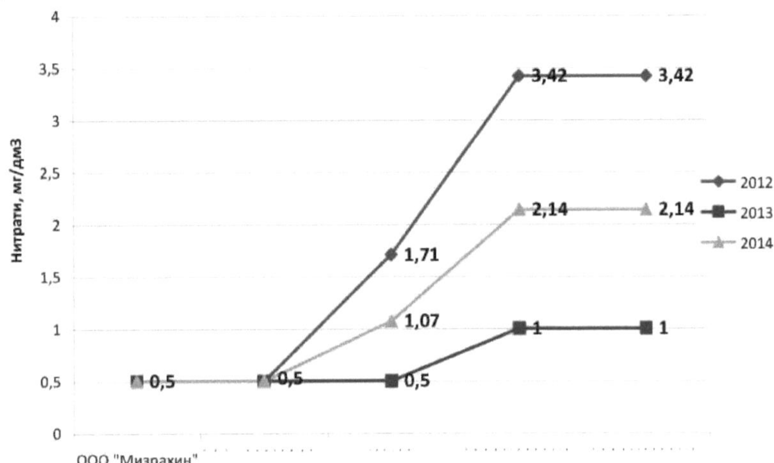


**Figure 23. Comparative characterization of tap water quality in Krivoy Rog district and pretreated drinking water from different manufacturers in terms of nitrate content.**

The increased oxidizability in all types of drinking water supply draws attention (Table 19).

*Table 19*

**Comparative characterization of quality indicators of tap drinking water in 1 rural taxon (Krivoy Rog district) and drinking water pretreated from different firms - producers on acidification, (mgO2/dm )$^3$**

| Years | Pretreated drinking water Mizrahin LLC | Pretreated drinking water LLC Anisimov | Tap drinking water in 1 taxon | Efficiency of drinking water pretreatment by Mizrahin LLC | Efficiency of additional treatment of drinking water of Anisimov LLC |
|---|---|---|---|---|---|
| 2012 | 1,62±0,01 | 1,27±0,20 | 5,57±0,08 | 3,44 | 4,38 |
| 2013 | 0,26±0,02 | 2,63±0,25 | 3,08±0,09 | 11,85 | 1,17 |
| 2014 | 3,70±0,10 | 3,77±0,02 | 4,04±0,83 | 1,09 | 1,07 |
| p | p = 0,1991 | | | | |

Note. 1p - level of significance of the efficiency of tap drinking water pretreatment from different firms - manufacturers according to the criterion $\chi^2$ - Pearson.

Thus, in pretreated water of the 1st producer (LLC "Mizrakhin") oxidizability exceeded MPC

2.2 times in 2012 and 5.0 times in 2014. Pretreated water of the 2nd producer (LLC "Anisimov") constantly exceeded the regulated value of oxidizability: 2.0 MAC - in 2012, 3.5 MAC - in 2013, 5.03 MAC - in 2014. The highest oxidizability was shown in tap water of taxon 1: 7.4 MAC - in 2012, 4.1 MAC - in 2013, 5.4 MAC - in 2014 ($p = 0.199$). According to GOST 7525:2014 [50] oxidizability should be no more than 0.75 mgO /$dm_2^3$ in water of decentralized drinking water supply. At the same time, the efficiency of water pretreatment increased in water of 1 producer: 3.44, 11.8 and 1.09 times; whereas in water pretreated by 2 producers: 4.38, 1.17 and 1.07 times.

This trend is probably due to the systematic inflow of organic matter into the source of water supply of taxon 1 - the Karachunovskoye reservoir, the water of which is used for drinking water supply of this taxon (Krivoy Rog district), as well as simultaneously used by specialized enterprises for additional treatment (firms - manufacturers LLC "Mizrakhin" and "Anisimov"), from tap drinking water coming through the system of centralized water supply in the Krivoy Rog urbanization zone, namely - Karachunovskoye reservoir (Fig. 24).

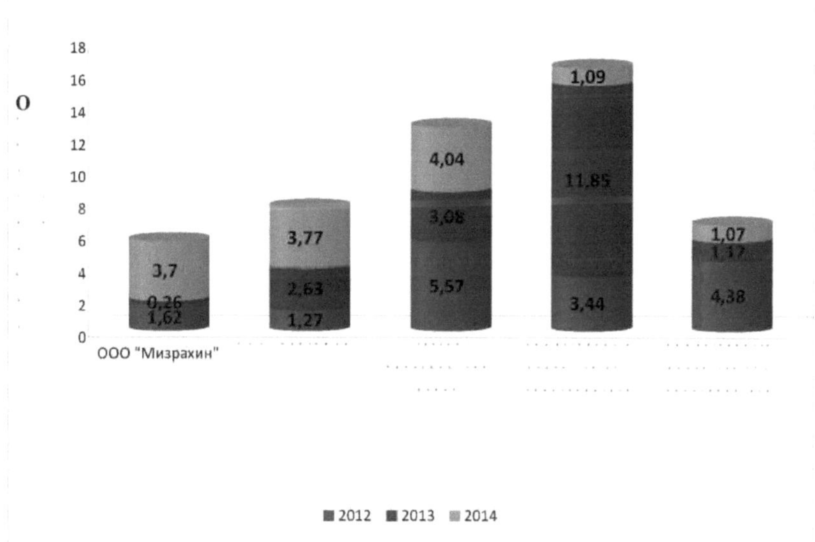


**Figure 24. Comparative characteristic of tap water quality in Krivoy Rog district and pretreated drinking water from different manufacturers in terms of oxidizability.**

## CONCLUSION

1. To date, the Ingulets River and the Karachunovskoye Reservoir have experienced intensive anthropogenic pollution associated with the activities of mining enterprises in the city of Krivoy Rog. Deterioration of water quality in the Ingulets River is a national problem. There is a risk of accumulation of significant volumes of highly mineralized water in the Karachunovskoye reservoir. Long-term discharge of mine, quarry, filtration and insufficiently treated industrial wastewater into the Ingulets River leads to a decrease in self-purification processes.

2. In addition, outdated drinking water treatment technologies do not perform a barrier function against many pollutants of natural water bodies, which correspond mainly to quality class 3, while water supply facilities are designed to effectively treat source water of quality class 1.

3. Improvement only of water treatment technologies in accordance with water classes of water source without ensuring proper sanitary and technical condition of water supply networks cannot contribute to obtaining drinking water of guaranteed high quality.

4. The structure of morbidity among the adult population in different taxa of Dnepropetrovsk region differs by classes of diseases. Thus, in taxon 1 the greatest specific weight is shown for diseases: X (27,9 %), IX (11,51 %), XIV (7,74 %), XIII (5,10 %) and XI classes (4,20 %); in taxon 2: For diseases of X (25.32 %), IX (13.9 %), XIV (8.19 %), XII (4.22 %), XIII (6.21 %) and IV classes (2.98 %); in taxon 3: for diseases of X (28.97 %), IX (13.55 %), XII (5.90 %), XIV classes (5.88 %) and XIII classes (4.01 %); in taxon 4: For diseases of X (26.17 %), IX (13.43 %), XIV (7.71 %), XIII (4.03 %) and XI classes (4.01 %); in 5 taxa: For diseases of X (27.79 %), IX (12.17 %), XIV (7.15 %), XIII (5.09 %), XII classes (4.44 %); in 6 taxon: for diseases of X (22.86 %), IX (13.71 %), XIV (6.84 %), XIII (6.26 %) and XI classes (4.26 %).

5. Thus, in the structure of all diseases among the adult population the pattern of the highest incidence of diseases of the respiratory system, circulatory system, genitourinary system, musculoskeletal system, digestive organs in all rural taxa of Dnipropetrovsk region was established. The lowest specific weight is established for diseases of XI class (K80- K87), XIV class (N25-N29) and (N17-N19), as well as XVII class, including congenital anomalies of the circulatory system in all taxa of the region.

6. The results of our study convincingly show that the largest specific weight in the structure of all diseases among the adult population, in all 6 types of taxa in Dnepropetrovsk region, caused by diseases of the respiratory, circulatory, digestive, genitourinary and bone and muscle systems and

other classes of diseases, correlates with the data of the literature [47, 48, 49]. In particular, infectious and parasitic diseases, diseases of the nervous system, blood and hematopoietic organs, including anemias, neoplasms, as well as some nosological forms - salt arthropathy, kidney and ureter stones, congenital anomalies (malformations), including circulatory system, occupy the last rank places in the structure of all diseases among rural residents in all taxa of the region for 2008 - 2013.

7. Comparative assessment of drinking water supply quality indicators in pretreated and tap water showed similarity in some indicators, such as: increased oxidizability, constant presence of ammonia nitrogen, which should be absent according to GOST 7525:2014 [50], similarity of salt composition (dry residue, chloride and sulfate content, pH), against the background of low concentration of TM (Cu, Zn, Mn), nitrites and nitrates, aluminum and fluoride in some years of observation in samples of pretreated drinking water from both manufacturers.

8. Such similarity of drinking water quality indicators of tap water and pretreated water in Krivorozhye urbanization zone is probably caused by simultaneous use as a water supply source - Karachunovskoye reservoir, water from which is used both for drinking water supply of 1 taxon (Krivorozhye district) and for water pretreatment by different firms - producers in the same Krivorozhye urbanization zone.

9. It was found that among the inhabitants of rural taxa of Dnepropetrovsk region the largest number of sources of drinking water supply is in 1 (244 water sources, i.e. 33.6 %), 6 (227, i.e. 31.3 %) and 5 taxa (107, i.e. 14.7 %); while the smallest number is in 4 (94, i.e. 13 %), 3 (33, i.e. 4.5 %) and 2 taxa (20, i.e. 2.7 %). At the same time, the largest number of decentralized sources of drinking water supply is in taxon 1 - 235 (43.6 %); the smallest is in taxon 3: 5 (0,9 %). Among centralized water sources, the highest number is in taxon 6: 79 (42.2 %), the smallest - in taxon 2: 20 (2.7 %). In all 6 taxa of Dnepropetrovsk region the total number of water supply sources is 725, among them: 187 - centralized, 538 - decentralized.

10. It was found that part of rural residents of the vast majority of rural taxa of Dnepropetrovsk oblast, which should be served by collective drinking water supply systems, do not have access to quality drinking water, because coverage rates in all taxa of the oblast by collective water supply systems were below the recommended "National target indicators" [420] from (18.5 - 1.5) to (25.9 - 2.0) times: (50 - 70) % - in villages, (90 - 100) % - in cities and towns.

11. The results of the conducted research allowed to scientifically substantiate a comprehensive approach to the improvement of the Ingulets River and Karachunovskoye reservoir - the main sources of centralized water supply to the rural population of Krivoy Rog urbanization zone; to form a set of measures aimed at the need for priority implementation of the monitoring system of health indicators

of the rural population; to outline the primary need for the use of pre-treated drinking water in rural taxons of Dnepropetrovsk region, which do not have access to drinking water in the rural areas of the Dnepropetrovsk region.

## REFERENCE LIST:


1. Serdyuk, A.M. 20 years of the National Academy of Medical Sciences of Ukraine: results and a look into the future / A.M. Serdyuk // Journal of the National Academy of Medical Sciences of Ukraine. - Vol. 19. - № 2. -2013. -C. 134 - 138.
2. Prokopov, V.A. State and quality of drinking water of centralized water supply systems in modern conditions (view of the problem from the standpoint of hygiene) / V.A. Prokopov // Hygiene of populated places. - Issue 64. - K., 2014. - C. 56 - 67.
3. Ryzhenko S.A. Ways to provide the population of Dnepropetrovsk region with quality drinking water / S.A. Ryzhenko, K.P. Vainer // Proceedings of the III International Scientific and Practical Conference "Healthy Lifestyle: Problems and Experience". - 2013. - C. 315 - 319.
4. Mokienko, A.V. Justification of research on the influence of water factor on the health of the population (literature review) / A.V. Mokienko, L.I. Kovalchuk // Hygiene of populated places. - Issue 64. - K., 2014. - C. 67 -76.
5. Gozhenko A.I. Water and health: an attempt to assess the problem: a review of the literature / A.I. Gozhenko, A.V. Mokienko, N.F. Petrenko // Health of Ukraine. - 2006. - C. 6 - 12.
6. Okrugin, Yu.A. Influence of microbiological and parasitological indicators of household sewage on water quality of open water bodies / Yu.A. Okrugin, S.V. Kapranov, L.I. Kosenko // Surrounding environment and health. - 2003. - № 4 (27). - C. 51 - 56.
7. Prokopov, V.A. Scientific and practical issues of providing the population of Ukraine with quality drinking water / V.A. Prokopov // Proceedings of the XIV Congress of Hygienists of Ukraine "Hygienic science and practice at the turn of the century". - T. 1. - Dnepropetrovsk, 2004. - C. 109 - 111.
8. Risk assessment of the development of non-carcinogenic effects on organs and systems of the population of single-industry towns and rural areas / V.M. Boev, D.A. Kryazhev, L.M. Tulina, A.A. Neplokhov, M.V. Boev // Proceedings of the Plenum of the Scientific Council of the Russian Federation on human ecology and environmental hygiene (December 11 - 12, 2014). - Moscow: FGBU "Research Institute of Human Ecology and Environmental Hygiene named after A.N. Sysin" of the Ministry of Health of Russia, 2014. -C. 55 - 57.
9. Onishchenko, G.G. On sanitary and epidemiological state of the environment / G.G. Onishchenko // Hygiene and sanitation. - 2013. - № 2. -C. 4 - 10.
10. Mudry I.V. Heavy metals in the environment and their effect on the body / I.V. Mudry, T.K.

Korolenko // Doctor's case. - 2002. -№ 5. -C. 6 -9.

11. Rukavichka, A.N. Organization of ecological and hygienic monitoring of heavy metals accumulation in the system "soil - vegetable production" on the territory of Dubrovitsky district of Rivne region / A.N. Rukavichka, I.V. Gushchuk // Hygiene of inhabited places. - Issue 62. - K., 2013. -C. 100 - 106.
12. Surveillance for waterborne - disease outbreaks / Boubetra L., Le Nestour F., Allaert C., Feinberg M. // Appl. Environ. Environ. Microbiol. - May 2011. -№ 77 (10). -P. 3360 - 3367.
13. Vulnerability of drinking - water wells / Parker A.A., Stivenson R.A., Raily P.L., Ombeki S.A., Komolleh C.L.. // Epidemiol. Infect. - October 2006. - № 134 (5). -P. 1029 - 1036.
14. Status of groundwater contamination in USA / Mausezahl D., Teller F., Iriarte M. // Clinical Microbiol. - July 2010. -№ 23 (3). -P. 507 - 528.
15. Water quality for cattle / Hattendorf J.L., Cattaneo M.D., Arnold V.F., Smith T.J. // Water Resources. - November 2010. -№ 49 (1). - P. 9 - 15.
16. Risk factors contributing to microbiological contamination of drinking water / Gueler F.M., Heiringhoff K.H., Engeli S.P., Heusser K.L.. // Environ. Health Perspectives. - October 2012. - № 6 (8). - P. 823 - 935.
17. Manual of social hygiene and health care organization in 2 volumes. T. 1 / Y. P. Lisitsyn, E. N. Shigan, I. S. Sluchanko [et al.]. Edited by Y. P. Lisitsyn. -M.: Medicine, 1987. - 432 c.
18. Prognostic assessment of morbidity indicators of the population living in the zone of influence of Khmelnitsky NPP / N. S. Polka, V. M. Dotsenko, A. I. Kostenko, I. V. Kakura // Proceedings of the XIX International Scientific and Practical Conference and Exhibition-Fair. Volume II. "Kazantip-EKO-2011", (June 6-10, 2011, AR Crimea, Cape Kazantip, Shchelkino). - Kharkov: UkrGSTC "Energostal", 2011. -C. 7-13.
19. Methodical recommendations "Assessment of public health risk from atmospheric air pollution" MR 2.2.12-142-2007. - Effective from 13.04.2007. - Kiev: Ministry of Health of Ukraine, 2007. - 39 c.
20. Chernichenko I. A. Scientific bases of hygienic rationing of chemical carcinogens at complex and combined intake into the organism : autorref. diss. doctor of medical sciences: spets. 14.02.01 "Hygiene" / I. A. Chernichenko. - Kiev, 1992. -44 c.
21. Trakhtenberg, I. M. Heavy metals as chemical pollutants of production and environment. Ecological and hygienic aspects / I. M. Trakhtenberg. - Minsk : Science and Technology, 1994. - 285 c.
22. Heavy metals in the environment and their effect on the organism (review) / R. S. Gildenskiold,

Y. V. Novikov, R. S. Khamidulin et al. // Hygiene and sanitation. - 1992. - № 5-6. - C. 6-9.
23. Yanysheva N. Ya. Hygienic problems of environmental protection from pollution by carcinogens / N. Ya. Yanysheva, I. S. Kireeva, I. A. Chernichenko et al. - Kiev: Zdorovye, 1985. - 102 c.
24. Persheguba Ya. V. Comparative assessment of carcinogenic risk of food and urban atmospheric air / J. V. Persheguba // Proceedings of the XIX International Scientific and Practical Conference and Exhibition-Fair. Volume II. "Kazantip-EKO-2011", (June 6-10, 2011, Crimea, Cape Kazantip, Shchelkino). - Kharkov: UkrGSTC "Energostal", 2011. - C. 19-23.
25. Hygienic assessment of water resources / V. L. Savina, S. V. Vitrischak, A. E. Akberov, V. V. Zhdanov // Proceedings of the XIX International Scientific and Practical Conference and Exhibition-Fair. Volume III. "Kazantip-EKO-2011", (June 6-10, 2011, AR Crimea, Cape Kazantip, Shchelkino). - Kharkov: UkrGSTC "Energostal", 2011. -C. 303-305.
26. Project "Dnipropetrovsk region. Territory Planning Scheme". Explanatory note. T. I, II / Ukrainian State Research Institute for Urban Design "Dnepropetrovsk". - Kiev. - 2009.
27. SanPiN No. 4630-88 Sanitary Rules and Norms for Protection of Surface Water from Pollution.
28. GOST 4808:2007 Sources of centralized drinking water supply. Hygienic and environmental requirements for water quality and sampling rules. - Kiev, 2012. - 27 c.
29. Hygienic requirements for drinking water intended for human consumption: State sanitary norms and rules GSanPiN 2.2.4-171-10; approved by Order of the Ministry of Health of 12.05.2010, № 40. - Access mode: http://normativ.ua/types/tdoc19074.php.
30. Health indicators of the population of Dnepropetrovsk region in2008-2013 . - Dnepropetrovsk: Main Department
health care of the regional state administration. Regional Center of Medical Statistics in Dnepropetrovsk, 2014. - 286 c.
31. ICD X: International Statistical Classification of Diseases and Related Health Problems. - 10th revision. - Geneva: WHO, 1995. -T. 1, Ч. 1. - 698 p., Ch. 2. -633 p., Ch. 2. -172 p.
32. Borovikov V. STATISTICA: The Art of Analyzing Data on the Computer. For professionals / V. Borovikov. - St. Petersburg, 2001. - 656 c.
33. Lapach S. N. Statistical methods in biomedical research using Excel / Lapach S. N., Chubenko A.. N., Chubenko A. V. V., Babich P. N.-K.: Morion, 2001. -408 c.
34. State of pollution of natural environment on the territory of Ukraine http://www.cgo.kiev.ua/index.pdf
35. Status of decentralized economic and

drinking water supply

Prokopov V.A., Kuzminets A.N., Sobol V.A. // Hygiene of inhabited places. - 2008. - Issue 51. - C. 63-68.

36. Ryzhenko, S.A. Trihalomethanes in drinking tap water / S.A. Ryzhenko // Preventive medicine. - 2009. - № 4. - C. 2021.

37. Koshelnik, M.A. Technogenic load on water bodies: consequences for public health / M.A. Koshelnik // Preventive Medicine. - 2009. -№ 4. - C. 28-31.

38. Water quality of centralized water supply in Ukraine on sanitary and microbiological indicators and associated infectious morbidity / Korchak G.I., Surmacheva A.V., Nekrasova L.S. et al. // Environment and Health. - 2012. - № 4. - C. 39-41.

39. From the experience of gossannadzor on the quality of packaged drinking water / Larchenko, V.I.; Ovchinnikova, V.A.; Zaitsev, V.V.; Ostapchuk, E.A.; Zadvornaya, V.V. // Environment and health. - 2008. - № 1 (44). - C. 43-44.

40. National Program for Ecological Improvement of the Dnieper Basin and Improvement of Drinking Water Quality. Resolution of the Verkhovna Rada of Ukraine of February 27, 1997.

41. Water as a source of infectious diseases / Nikolenko P. P. P., Beloivanenko V. I., Kuleshov N. I. // Med. Vesti. - 1997. - № 3. - C. 14-16.

42. Influence of microbiological and parasitological indicators of domestic wastewater on the water quality of open water bodies / Okrugin Y. A., Kapranov S. V., Kosenko L. I. and others // Environment and Health. V., Kosenko L. I. and others // Environment and Health. - 2003. - № 4 (27). - C. 51-56.

43. Alekseenko, N.N. Ecological assessment of water quality condition of the Kremenchug reservoir / N.N. Alekseenko // Environment and Health. - 2004. - № 2 (29). - C. 30-35.

44. Palchitsky A. M. Kakhovka reservoir: Current state and possible ecological and sanitary forecast /

A.M. Palchitsky // Hygiene and sanitation. - 1991. -№ 10. -C. 21-25.

45. Ryzhenko, S.A. Some aspects of the state of the environment of the technogenic region and approaches in the organization of the work of the state epidemiological service of the Dnepropetrovsk region / S.A. Ryzhenko // Environment and Health. - 2004. - № 2 (29). - C. 48-53.

46. Hryhorenko LV. Potable water quality in the Karachunyvskyi reservoir / L.V. Hryhorenko // Austrian Journal of Technical and Natural Sciences. - 2014 (February 28). -№1. -C.40 -45.

47. Scientific and methodological approaches to the calculation of actual and prevented medical,

demographic and economic losses associated with the negative impact of environmental factors / N. V. Zaitseva, I. V. May, D. A. Kiryanov // Proceedings of the Plenum of the Scientific Council on Human Ecology and Environmental Health (December 11 - 12, 2014). - Moscow: FGBU "Research Institute of Ecology and Environmental Health named after A. N. Sysin of the Ministry of Health of the Russian Federation". - C. 85 - 103.

48. Relation of chronic non-infectious diseases with the state of the environment / Yu.A. Rakhmanin, A.A. Stehin, G.V. Yakovleva, VV. Ryabikov // Proceedings of the Plenum of the Scientific Council on Human Ecology and Environmental Hygiene (December 11 - 12, 2014). - Moscow: FGBU "Research Institute of Ecology and Environmental Health named after A. N. Sysin of the Ministry of Health of the Russian Federation". - C. 78 - 93.

49. Analytical problems in the study of the complex effect of environmental factors on the health of the population / AG Malysheva, E. G. Rastiannikov, N. Yu. Kozlova // Proceedings of the Plenum of the Scientific Council on Human Ecology and Environmental Hygiene (December 11 - 12, 2014). - Moscow: FGBU "Research Institute of Ecology and Hygiene named after A. N. Sysin of the Ministry of Health of the Russian Federation". - C. 118 - 140.

50. Drinking water. Requirements and methods of quality control. GOST 7525:2014. - Kiev: Ministry of Economic Development of Ukraine, 2014. - 25 c.

**Grigorenko Lyubov Viktorovna**, Candidate of Medical Sciences, Associate Professor of the Department of Hygiene and Ecology "Dnepropetrovsk Medical Academy of MHI". Second higher education in the direction of training 6.020303 "Specialist in Philology. Translator of the English language". Conducts practical classes and consultations, gives lectures on the subject: "General hygiene and ecology" at English-speaking foreign students and students of medical faculties of VI courses in the specialty: "Medicine". Author of 130 publications: 79 of scientific and 51 educational and methodological character, including 17 in fakh publications. After the defense of the candidate's thesis she published 102 scientific articles: 59 - in scientific journals and 43 educational-methodical, including 14 works in fakh publications, 10 - foreign articles, 4 - in international scientific-metric journals; 10 teaching aids for English-speaking students; 6 author's certificates.

Member of the Federation of the National Team of Scientists of the IASHE international project (in London). Three times she was awarded a bronze medal for the best publication in English as a prize-winner of the I, II and III stages of competitions in the branch "Medicine and Pharmacy, Biology, Veterinary Medicine and Agriculture", section: "Hygiene".

Printed by Books on Demand GmbH, Norderstedt / Germany